T0339902

STREAMLINE NUMERICAL WELL TEST INTERPRETATION

Theory and Method

STREAMLINE NUMERICAL WELL TEST INTERPRETATION

Theory and Method

YAO JUN
WU MINGLU

ELSEVIER

AMSTERDAM • BOSTON • HEIDELBERG • LONDON
NEW YORK • OXFORD • PARIS • SAN DIEGO
SAN FRANCISCO • SINGAPORE • SYDNEY • TOKYO
Gulf Professional Publishing is an imprint of Elsevier

Gulf Professional Publishing is an imprint of Elsevier
225 Wyman Street, Waltham, MA 02451, USA
The Boulevard, Langford Lane, Oxford OX5 1GB

First edition 2011

Copyright © 2011 Yao Jun and Wu Minglu. Published by Elsevier Inc. All rights reserved

The right of Yao Jun and Wu Minglu to be identified as the authors of this work has been asserted
with the Copyright, Designs and Patents Act 1988.

No part of this publication may be reproduced, stored in a retrieval system or transmitted in
any form or by any means electronic, mechanical, photocopying, recording or otherwise
without the prior written permission of the publisher

Permissions may be sought directly from Elsevier's Science & Technology Rights
Department in Oxford, UK: phone (+44) (0) 1865 843830; fax (+44) (0) 1865 853333;
email: permissions@elsevier.com. Alternatively visit Science and Technology web site at
www.elsevierdirect.com/rights for futher information.

Notice
No responsibility is assumed by the publisher for any injury and/or damage to persons or
property as a matter of products liability, negligence or otherwise, or from any use or
operation of any methods, products, instructions or ideas contained in the material herein.
Because of rapid advances in the medical sciences, in particular, independent verification
of diagnoses and drug dosages should be made.

Library of Congress Catalogue in Publication Data
A Catalogue record for this book is available from Library of Congress

British Library Cataloguing in Publication Data
A Catalogue record for this book is available from the British Library

ISBN: 978-0-12-810374-6

For information on all Academic Press publications visit
our website at *elsevierdirect.com*

Typeset by: Thomson Digital, India

Printed and bound in USA

11 12 13 11 10 9 8 7 6 5 4 3 2 1

Working together to grow
libraries in developing countries

www.elsevier.com | www.bookaid.org | www.sabre.org

ELSEVIER   BOOK AID   Sabre Foundation
            International

Contents

7. Streamline Numerical Well Testing Interpretation Model Considering Components

8. Streamline Numerical Well Testing Interpretation Model for a Multi-Layer Reservoir in Double-Porosity Media

9. Streamline Numerical Well Testing Interpretation Model of a Horizontal Well

10. Multi-parameter Streamline Numerical Well Testing Interpretation Method

11. Software Programming of Streamline Numerical Well Testing Interpretation

12. Field Application of Streamline Numerical Well Testing Interpretation Software

Foreword

This work, *Streamline Numerical Well Test Interpretation Theory and Method*, written by Professor Yao Jun, is based on the summary and abstraction of scientific achievements and application experiences after years of data accumulation in the numerical well test field. It gives expression to new trends in theories and methods in this field with distinct originality and practicability, provides an effective method for the application of well testing data in reservoir fine description, especially the determination of residual oil distribution. Also, the way in which the theories are systematically described and the integrity of this book provide a factual basis for readers to understand streamline numerical well test interpretation theory and method in detail.

In this book, the following principles are outlined, in terms of theories and methods:

1. Strong innovations. The streamline method is used in numerical well test interpretation for the first time, which breaks the traditional idea that one well test equation should be built for one testing well. Based on a true reservoir model and taking into consideration the influence of production history, oil layer heterogeneity, multi-phase fluid and non-uniform distribution, multi-well interference, layer cross-flow and complex reservoir boundaries, the streamline numerical well test interpretation model is built including a mathematical model of production and testing periods. The mathematical model of the production period is used to simulate production history, and the streamline method is used to quickly calculate the distribution of pressure, saturation and streamlining; the mathematical model of the testing period is built by flow equations of each streamline surrounding testing well, which is used to simulate bottom hole pressure change of testing period, to obtain theoretical pressure response of the testing well and to provide a detailed description of formation and flow.

2. Strong systematicity. Streamline numerical well test interpretation theory and method systems are formed with all kinds of numerical well test interpretation models from single-layer reservoir models to multi-layer reservoir models: the water-flooding reservoir model to the chemical-flooding (polymer flooding, alkaline flooding, alkaline/polymer binary combination flooding) model; the sandstone reservoir model to the carbonate dual porosity media reservoir model; the fully perforated well model to the partially perforated well model; the regularly damaged well model to the irregularly damaged well model; the vertical testing well model to the horizontal testing well model.

3. Strong practicability. The streamline numerical well test interpretation method avoids the disadvantages of conventional well test and numerical well test interpretation methods; this well test interpretation model considers the influence of complex factors including development and geology, which coincides better with real reservoirs than other models. With powerful analysis capabilities and reliable results, this method can not only provide conventional well test parameter information, but can also provide the dynamic parameter distribution of residual oil and polymer concentrations. In addition, the streamline method and the multi-population genetic algorithm greatly improve the speed of well test interpretation and application scale with strong field practicability. The streamline numerical well test theory and method introduced in this book has been programed to streamline numerical well test interpretation software with complete functionality, reliable practicability and independent intellectual property rights, which is also widely used in Shengli, Zhongyuan, Nanyang and Dagang Oilfields and provides a considerable economic benefit.

With a series of pictures, and with full and accurate data, this book not only has a high academic value, but also has a broad application, and can serve as a reading and reference book for scientists, engineers and college students in petroleum and other relevant fields.

Academician of the Chinese Academy of Science
Guo Shangping

Correct evaluation of dynamic parameters in the developed reservoir, especially residual oil distribution, is the basis for establishing stimulation measures or enhanced oil recovery (EOR) schemes scientifically and reasonably. Also, determination of storage parameters and residual oil distribution with well testing data is a convenient, economical, reliable and practical method.

The conventional and modern well test interpretation method, which is based on a reservoir conceptual model and analytical solutions, is a relatively practical method in oil field exploration and during the early development period. However, most oil fields at home and abroad enter the middle- and later-development periods, and reservoir flow environment becomes increasingly complex (e.g., formation heterogeneity, fluid non-uniform distribution, multi-well interference and cross-flow between layers). This situation is totally different with the ideal models of conventional and modern well testing. In this case, conventional well test and modern well test interpretation methods could not satisfy the needs of oil field development and evaluation.

In order to meet the demands of testing data interpretation in middle and later periods, the numerical well test interpretation method was proposed in the 1990s. This method is based on a true reservoir model, builds the well test interpretation model by considering complex boundaries, production history, multi-phase flow, heterogeneity well pattern and well type, solves with numerical methods and interprets with automatic matching methods for multi-parameter well testing. In order to distinguish it from the well test interpretation method based on analytical solutions, this method is called the ''numerical well test interpretation method''. At present, the main solution methods for the numerical well test interpretation model include finite difference, finite element, boundary element and Green element methods, which are all based on 2-D or 3-D grid generation to realize pressure dynamic fine simulation of the testing well by well grid refinement; the speed of calculation and the accuracy do not easily satisfy the needs of numerical well test interpretation. Hence, the use of present numerical well test interpretation methods is poor and it can not be widely applied.

For the problem of slow speed and small application scale during the numerical well test interpretation method with finite difference algorithm, the streamline method is introduced into well test interpretation and the streamline numerical well test interpretation method is proposed. The mathematical model adopted in this method is divided into production and testing periods. The mathematical model of the production period is used to simulate production history, and the streamline method is introduced for fast calculation of pressure distribution, saturation distribution and concentration distribution (chemical flooding), which are taken as the initial condition of the testing well in the testing period. Meanwhile, the mathematical model of the testing period is used to simulate bottom-hole pressure change of the testing well in the testing period, and it is made by the flow equations of each streamline surrounding testing well, then a simultaneous solution is used to obtain theoretical pressure response. The models in the two periods are the true reservoir model, which can consider the influence of complex factors, such as production history, oil layer heterogeneity, multi-phase fluid and non-uniform distribution, multi-well interference, cross-flow between layers and complex reservoir boundary; furthermore, the introduction of the streamline method guarantees the calculation speed and accuracy. By changing the parameters of the well test interpretation model constantly and by automatic matching of theoretical pressure response and testing pressure data of the testing well in the well test interpretation model, accurate well test interpretation parameters can be obtained.

This method has formed a consummate well test theoretical system and interpretation method from single-layer reservoirs to multi-layer reservoirs, sandstone reservoirs to fractured dual-porosity media reservoirs, water flooding reservoirs to polymer, alkaline and chemical combination flooding reservoirs, which greatly enriches numerical well test interpretation methods. Meanwhile, streamline numerical well test interpretation software, with independent intellectual property rights, has been programed which is widely applied in Shengli, Zhongyuan, Henan and Dagang oil fields, and a practical method to determine the distribution of permeability and residual oil has been established based on well test data.

This book presents the research achievements in this field in the last 10 years in 12 chapters. Chapter 1 describes the development history of well test theory, analyzes the limitations of modern well test interpretation methods, and then proposes the concept and framework of numerical well testing. Chapter 2 introduces basic principles and solution procedures of streamline numerical simulation theory and method, which will help readers who have not previously used the streamline numerical simulation method. Chapters 3–9 study streamline numerical well test interpretation models in many kinds of reservoirs and wells systematically: from single-layer reservoirs to multi-layer reservoirs; single-layer sandstone water flooding reservoirs to multi-layer sandstone water flooding reservoirs; multi-layer sandstone water flooding reservoirs to multi-layer sandstone chemical flooding models and the model considering components; single-porosity media reservoir to dual-porosity media reservoir; normal inner boundary conditions with a totally perforated oil layer to complex near well boundary conditions with a partially perforated oil layer considering perforation location and irregular damage; conventional well (straight well) to complex structural well (horizontal well). In particular, the numerical well test interpretation method is firstly introduced to chemical flooding and dual porosity media reservoirs, which enriches and develops numerical well test interpretation methods and builds a better methodology system. Chapter 10 presents a multi-parameter streamline numerical well test automatic match interpretation method based on a double-population genetic algorithm, which lays the foundations to fast automatic match of numerical well testing. Chapter 11 introduces streamline numerical well test interpretation software, with independent intellectual property rights, which is programed based on the above theoretical studies. Chapter 12 describes the application study of streamline numerical well test software, and the programed software is applied in actual fields with many different types of reservoir. The biggest application scale could consider 177 wells (121 production wells, 56 injection wells) working at the same time, while the longest simulation history is 35 years and the most simulation layers is 5 layers. Also, application reservoir types refer to water flooding reservoirs, polymer flooding reservoirs and alkaline–polymer combination flooding reservoirs. The interpretation results include not only conventional well test interpretation parameters (well bore storage coefficient, skin factor, etc.), but also permeability distribution of whole reservoirs, residual oil distribution, chemical concentration distribution (chemical flooding) and displacement front position, etc.

Chapter 1

Numerical Well Testing Interpretation Theory and Method

1.1. WELL TESTING OVERVIEW

In order to obtain maximum development benefits, a practical reservoir model which conforms to real reservoir conditions needs to be built. Using reservoir models and reservoir engineering methods, varied oil and gas field development plans and operation modes can be simulated to predict precisely the dynamic characteristics of the reservoir and wells; thus scientific and suitable development decisions can be made. Building a reservoir model needs geological data, geophysical data, logging data, core analysis data and production performance data, all of which can be obtained by direct measurement such as core and reservoir fluid sampling and data interpretation such as analysis of seismic data, logging data, well testing data and pressure, volume, temperature (PVT) data. Seismic data, logging data and core analysis data can only provide a static description of the reservoir, while well testing data, serving as the main foundation of reservoir model building, can supply dynamic information about reservoirs and wells.

Through well test analysis, we can obtain various dynamic data of reservoirs and wells, such as effective permeability (formation capacity, flow coefficient), initial or average reservoir pressure, damage or improvement conditions for near the well bore area, producing reserves, fault and boundary conditions, inter-well communication situations, and so on.

At present, well test technology is widely used in oil and gas fields and has become one of the principle technologies employed during oil and gas field exploration and development. Nowadays, many waterflood oil fields have entered the late development period, and current well test theory and interpretation methods can not satisfy the actual production requirements. The existing problems in well testing are discussed below by reviewing and analyzing the well test development history and the present situation.

1.2. DEVELOPMENT HISTORY OF WELL TESTING THEORY

Well testing is an important part of oil and gas reservoir monitoring, which refers to various areas including reservoir geology, reservoir physics, formation and

Streamline Numerical Well Test Interpretation. DOI: 10.1016/B978-0-12-386027-9.00001-0
Copyright © 2011 by Elsevier Ltd

fluid properties, flow theory, optimization theory, computer technology, testing techniques, measuring instruments, and so on. Well testing theory has developed following the development of testing instruments.

1.2.1. Development History of Testing Instruments

More than half a century ago, we could only use the recording pen to record the maximum bottom-hole pressure using a simple glass tube manometer. After many years of development, the design and manufacture of manometers have become very sophisticated and greatly improved. Mechanical manometers comprising three key components, which include recording systems, travel-time systems and pressure sense systems, can record various characteristics of bottom-hole pressure changes. The manometers can work for 360–480 h down hole and endure temperatures of between 150 and 370 °C. The accuracy achieved can be to within 0.2% and there are dozens of varieties.

Over the past 40 years, computer technology has developed rapidly and has been used to advance the field of well testing. In the late 1960s, Hewlett–Packard Company successfully developed the first quartz electronic manometer with an accuracy of within 0.025%. Its degree of sensitivity is up to 0.00014 MPa and the scan rate can reach one measuring point per second. Quartz electronic manometers can be controlled remotely, bottom-hole pressure change can be observed through a secondary instrument and the length of measuring time can be adjusted according to requirements, thus it can greatly improve the quality of the well test data and the effectiveness of data analysis. Currently there are dozens of kinds of electronic manometers; some can directly read bottom-hole pressure and temperature while others can store the recorded data underground and then the data can be played back when the instrument is retrieved. The quartz electronic manometer is currently one of the most precise and sensitive types of manometers. The emergence of electronic manometers with high accuracy further promotes the development of well test theory, enhances the reliability of well test interpretation results and enlarges the application area of well test technology.

1.2.2. Development History of Well Testing Theory and Interpretation Methods

The development history of well test theory and interpretation methods can be divided into two stages, as described below.

1.2.2.1. Conventional Well Testing Analysis Methods Before the 1970s

Before the 1940s, our knowledge was confined to static pressure because the manometers could only measure the reservoir static pressure. Later, it was found that the measurement of static pressure was related to time, and the length of pressure build-up reflected the formation permeability around the well bore.

In 1933, Moore et al. published a paper which proposed a method to determine formation permeability using dynamic pressure data. Then two papers published in 1950 laid the foundation for well test theory. One was published by Horner (1950), who proposed using a diagram method to interpret pressure test data, i.e. he proposed that there is a linear relationship between the pressure build-up value and the logarithm value of Horner time (Horner Method). The other paper, written by Miller et al. (1950), proposed the linear relationship between pressure build-up value and shut-in time (MDH Method). Both of these two methods are used currently. Generally speaking, the Horner method is applicable to new wells in oil fields which are not fully developed, while the MDH method fits wells in oil fields which have been developed for a period of time. These two methods are the representatives of conventional well test interpretation methods. The so-called conventional well test interpretation methods refer to the methods represented by the Horner method which use the slope and intercept of the straight line to reversely solve the formation parameters. In addition, conventional well test interpretation methods also include the MBH method, the Y-function method, and so on.

Conventional well test methods mainly focus on the interpretation of the test data of isotropic and homogeneous reservoirs with comparatively mature theories, simple principles and convenient applications. However, conventional well test interpretation methods also present several drawbacks.

1. Conventional well test interpretation methods mainly focus on the interpretation of mid- to late-period pressure data, which needs a long period of well test time, thereby affecting oil production. Furthermore, it is rather difficult to get mid- to late-period well test data for reservoirs with very low permeability.
2. While using the conventional well test interpretation methods, the selection of the straight-line section will affect the subsequent interpretation of results. This artificial selection will inevitably produce some error.
3. Conventional well test interpretation methods do not make a detailed analysis of early period data, so they can not precisely evaluate the well bore storage capacity.
4. The limited date obtained during analysis with conventional interpretation methods brings some difficulties in recognition of reservoir model. Sometimes one single curve shape reflects characteristics of different reservoir models.

1.2.2.2. Modern Well Testing Analysis Methods After the 1970s

The modern well test interpretation methods developed since the late 1960s solve the above problems to some extent. The fundamental principle of these methods is to rebuild the mathematical model with various boundary conditions based on the original reservoir model, then to solve the mathematical model by analytical or semi-analytical methods and draw well test interpretation charts for analysis. For instance, Ramey (1970) delivered a type-curve chart of a well with well bore storage effects and skin effects in an infinite homogeneous reservoir. Then Gringarten and Ramey (1973) adjusted this chart to make it more applicable and to reduce the ambiguity of interpretation. In 1974 Gringarten et al. published a paper which delivered type-curves for wells with vertical fractures. In 1980, Bourdet et al. (1980) presented type-curves of dual porosity reservoirs and the corresponding pressure derivative curves, which made the model and flow regime diagnosis more precise. The introduction of systematic analysis methods further developed the well test interpretation method, which evolved from type-curves matching to various automatic matching methods such as non-linear regression, neural networks, control theory and so on. Based on these theories and with the assistance of computers, a large amount of practical well test interpretation programs were designed and widely used in oil and gas fields.

Modern well test interpretation methods have the following characteristics:

1. The application of systematic analysis concept and numerical simulation method further developed the well test interpretation theory.
2. With consideration for the impact of well bore storage effects and skin factors on pressure response, modern well test interpretation methods can analyze the early data and obtain more valuable information from it, which could not be utilized in the past.
3. It improved the conventional well test interpretation methods by giving the approximate beginning time of the semi-log straight line, which enhanced the reliability of semi-log curve analysis.
4. Through the matching of actual pressure data and dimensionless pressure–dimensionless time curves in theoretical charts, reservoir parameters can be quantitatively analyzed partially or generally and some parameters which could not be calculated using conventional well test interpretation methods now can be obtained.
5. Different reservoir models can be recognized using pressure derivative curves, which provide the basis for intended analysis and enhance the accuracy of the analysis.
6. The entire interpretation process is interpreting-while-verifying. The recognition of each flow stage and the computation of each parameter can almost be obtained from two different sources and then be compared.
7. The final interpretation of results can be verified and matched using numerical simulation, which enhances the reliability and accuracy of the interpretation of results.

1.3. LIMITATIONS OF MODERN WELL TESTING INTERPRETATION METHODS

Since the 1970s, well test theory and interpretation methods have not made any big progress in essence, which means that well test theory and interpretation methods can not meet the requirements of oil and gas fields' fast development. The commonly used Gringarten type-curve diagram well test interpretation method is used as an example below to illustrate the limitations of modern well testing interpretation methods.

1.3.1. Well Testing Interpretation Model of Homogeneous Reservoir

The Gringarten type-curve chart is currently a widely used modern well test interpretation chart which is delivered by solving well test interpretation models of homogeneous reservoirs with the following assumptions.

1. The reservoir is composed of infinite, homogeneous and isotropic single oil layer (i.e. the permeability, porosity, thickness and fluid properties are all uniformly distributed). The testing well is located in the reservoir center.
2. Single-phase flow underground with constant production rate.
3. Fluid is slightly compressible and the compressibility is a constant.
4. The well produces at a constant rate and the formation pressure is uniformly distributed before production.
5. The model neglects gravity and capillary pressure effects.

Based on the above assumptions, the physical model is shown in Fig. 1.1. The well testing interpretation model is

$$
\begin{cases}
\dfrac{\partial^2 p_D}{\partial r_D^2} + \dfrac{1}{r_D}\dfrac{\partial p_D}{\partial r_D} = \dfrac{\partial p_D}{\partial t_D} \\[2mm]
p_D(r_d, t_D)\big|_{t_D \pounds 1/2\,0} = 0 \\[2mm]
C_D \dfrac{dp_{WD}}{dt_D} - \left(\dfrac{\partial p_D}{\partial r_D}\right)_{r_D = 1} = 1 \\[2mm]
p_{WD} = \left[p_D - s\left(\dfrac{\partial p_D}{\partial r_D}\right)\right]_{r_D = 1} \\[2mm]
\lim_{r_D \to \infty} p_D(r_D, t_D) = 0
\end{cases}
$$

where, subscript D represents dimensionless. p_D is dimensionless pressure; r_D is dimensionless radius; t_D is dimensionless time; C_D is dimensionless well bore storage factor; p_{WD} is dimensionless bottom-hole flow pressure; S is skin factor.

The mathematical model above can be solved using an analytic method (through analytical conversion of the Laplace domain solution), a semi-analytic

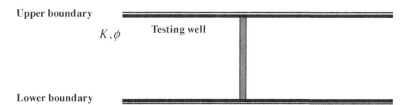

FIGURE 1.1   Schematic map of homogeneous reservoir model with a single well.

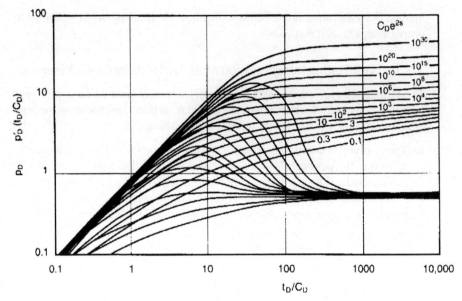

FIGURE 1.2   Modern well test interpretation type curves of a homogeneous reservoir.

method (through numerical conversion of the Laplace domain solution), and a numerical solution (through the finite difference method). Then, by the combination of these parameters, we can obtain the dimensionless pressure and derivative type-curve chart (as shown in Fig. 1.2).

1.3.2. Limitations of Well Testing Interpretation Models

From the model assumptions we know that the well test interpretation above is based on a comparatively ideal reservoir model (as shown in Fig. 1.3). It is very different from the real reservoir model because it does not consider many geological and developmental factors which can affect the pressure response of the testing well, including:

1. Reservoir heterogeneity. In the horizontal direction, it includes the heterogeneous distribution of reservoir rock properties such as permeability, porosity and thickness, and fluid properties such as fluid density, oil/gas/water saturation and compressibility. In the vertical direction, the reservoir is composed of several layers and each layer's lithology and fluid properties are different because of the sedimentary effect. These heterogeneities result in the existence of residual oil and low oil field recovery. However, the well test interpretation model above does not consider these factors.

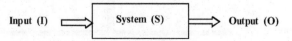

FIGURE 1.3   Schematic diagram of systematic analysis.

2. Impact of multi-well production on well test. In real reservoir conditions, production wells or injection wells may exist around the testing well and these wells can dramatically affect the fluid distribution. If these wells around the testing well do not stop production or injection during the testing, they will certainly affect the testing well. In addition, even if these wells are shut in, the irregular oil/water boundary formed around the testing well will also affect the pressure response of the testing well and this effect can not be substituted by simple linear boundary or fault.

3. Multi-phase flow effect. Currently, common well test interpretation software all use the Perrine–Martin equation to simplify multi-phase flow testing. This method only adjusts the parameters such as compressibility coefficient and fluid mobility and its flow equation form is identical to that of single-phase flow. Furthermore, underground multi-phase fluids can cause saturation distribution and phase change such as condensation and miscible phase, and then cause the corresponding change in fluid properties underground such as density and compressibility factor, which will definitely affect the pressure response of the testing well.

4. Impact of production history. At present, commonly used well test interpretation models all make the assumption that the reservoir pressure and fluid saturation are all uniformly distributed before the well test. Actually, production or injection history will cause redistribution of underground fluids and pressure, and this non-uniform distribution before well test will definitely affect the pressure response of the testing well.

Due to the big difference between the real reservoir model and the ideal reservoir concept model on which the well test interpretation base, the well test interpretation theory and methods are apparently deficient and their field applications are seriously restricted. It can be said that present well test theory and interpretation methods to some extent can not satisfy the requirements of oil field production tail and the development of some special oil fields.

With the development of oil/gas exploration and development techniques, more and more types of reservoirs, which also include some tough-to-tap reservoirs, are put into production. According to the underground fluid properties, these reservoirs can be classified as heavy oil reservoirs, black oil reservoirs, volatile reservoirs, condensate gas reservoirs, gas reservoirs, and so on. On the basis of permeability, reservoirs can be classified as low-permeability reservoirs and high-permeability reservoirs. In terms of development mechanisms, there are elastic developments, water flooding developments, gas injection developments (miscible or immiscible), chemical flooding developments, and so on. Each kind of reservoir has its own special development methods and unique underground flow mechanism. So the corresponding well test theory and interpretation methods for these reservoirs are much more complicated. Reservoir type, properties and drive mechanism must be taken into consideration when computing the pressure response.

Although interpretation software makes some adjustments for the inner and outer boundary and variant production rate of the model, the well test interpretation model is not changed in essence. Consequently, much field information, especially the test data of production tail, are hard to match. Some test curves are well matched because man-made boundaries or faults are added, but actually there is no boundary and it is caused by the oil/water distribution. The differences between the shapes of some real test curves and type-curves is entirely a reflection of the complexity of underground reservoir. This ideal and simplified reservoir concept model can not represent the underground reality. As a result, currently used over-ideal well test theory apparently can not satisfy the practical requirements of oil and gas field development.

1.3.3. Cause Analysis

There are many causes of the limitations listed above. Firstly, due to the restricted ability to solve the mathematical model, it is hard to develop so-called well test interpretation type-curve charts and interpret the real test data if the actual flow model is not simplified. Secondly, in the past few decades, well-known experts in the field of well testing almost all focus on the well test itself and do not pay enough attention to the combination with reservoir geology and reservoir engineering methods. In fact, well test theory and interpretation methods refer to many scientific domains and interdisciplinary combination is needed to promote the development of this subject. With the combination of reservoir engineering and well test interpretation methods, we examine the essence of well testing from different perspectives, which will be very beneficial to the further development of well test theory and interpretation methods.

1.4. ESSENCE OF WELL TESTING INTERPRETATION

1.4.1. Perspective from Systematic Analysis

Any study object can be seen as a System (S). If a stimulation or called Input (I) is imposed on the system, it will give a corresponding reflection called Output (O), as Fig. 1.3 shows.

There are two categories of problems in systematic analysis. One is where the system structure and input signal I are known, unknown output O is to be solved. This type of problem is called Direct Problem, indicated by

$$I \times S \to O$$

The other is where the system S is unknown while input I and output O of the system are known, the structure or characteristic of the system S is to be solved reversely, indicated by

$$O/I \to S$$

If the reservoir and the testing well are seen as one system, an input signal I is imposed on S during the test, i.e. the testing well is opened and produces

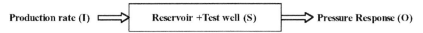

FIGURE 1.4   Schematic diagram of well test.

(or injects) at a constant rate or is shut in, then it will cause the pressure change in the system S, which is the output signal O of S and can be measured by test instruments, as Fig. 1.4 shows.

Well testing is the process of measuring the changes of production or injection rate and the bottom-hole pressure of the testing well, i.e. obtaining input or output signals of the system. While the purpose of well test interpretation is to recognize the system S, namely determining the type and characteristic parameters of the reservoir by using these data including input I (production rate) and output O (pressure change) with the combination of actual initial and boundary conditions of the reservoir and other relevant information of the reservoir and wells. That is to say, the process of well test interpretation is a typical inverse problem.

There are many methods to solve an inverse problem. To make it convenient for field use, well test researchers developed well test interpretation type-curves and the mission of well test interpretation can be finished by matching the real test curves with type-curves to determine the characteristic parameters of the reservoir and testing well. Some researchers proposed automatic matching methods and ascribed the well test interpretation to an optimization problem, which can be described as follows:

$$V\min = \sum_{i=1}^{N} \left(p_{\text{model }i} - p_{\text{real }i}\right)^2$$

$$V\min = \sum_{i=1}^{N} \left|p_{\text{model }i} - p_{\text{real }i}\right|$$

There are also many methods to solve this type of problem, such as the optimization method, the neural network method, the cybernetics method, the currently used generalized pulse spectrum technique and perturbation method, and so on.

The introduction of systematic analysis methods into well test interpretation is great progress which makes well test interpretation results more sophisticated and suitable to reality.

1.4.2. Perspective from Reservoir Model Building

The purpose of well test interpretation is to determine reservoir type and its characteristic parameters which can compose a simplified reservoir model. Whether this reservoir model fits the practical condition or not is determined by whether or not the simplification of the well test interpretation model is reasonable.

Well test interpretation is an adjustment or verification process of the reservoir model and this concept is critical to the development of well test theory.

Starting from this concept, we can deepen our understanding of well testing and enlarge the application range of well testing in oil and gas field development. Apparently, if the well test interpretation model is over-simplified, the reservoir geological model obtained will also be over-simplified.

The present well test interpretation model is a single-phase, single-well, single-layer and homogeneous (including rock properties, fluid properties and fluid distribution) simplified model, so the reservoir model determined by these well test theories and interpretation methods is certainly a single-phase, single-well, single-layer and homogeneous simplified model and the parameters determined by well test interpretation can only reflect the average properties of the reservoir which are of little value to the fine dynamic performance evaluation in oil field production tail. Clearly, this over-simplified model can not reflect the actual complex reservoir model very well. From this perspective, the well test interpretation model is very important and must be identical to the real condition. Only by satisfying this requirement can we determine the parameters of reservoirs and testing wells exactly.

Whether we can get finer and more microscopic (not average) geological characteristics from well test interpretation is completely determined by whether or not the well test interpretation model has considered these factors. Currently proposed methods using single-well test interpretation models to determine the distribution of residual oil saturation are completely unnecessary because the well test interpretation model does not take the factors which affect the distribution of residual oil into consideration.

From the perspective of reservoir model building, the well test interpretation model must be much closer to the real reservoir geological model and has to consider the following geological and developmental factors:

1. Reservoir heterogeneities in horizontal or vertical directions.
2. Impact of multi-well production on the testing well.
3. Effect of multi-phase fluid distribution and their phase changes.
4. Impact of production history on the pressure response of the testing well.

That is to say, the constructed well test interpretation model must be the real reservoir model; only in this way can we adjust or validate reservoir models through well test data analysis and finally obtain the real parameters of reservoir and the testing well.

1.5. BRIEF INTRODUCTION TO THE NUMERICAL WELL TESTING METHOD

Through the analysis above we know that to consider the effects of more factors on the pressure response of the testing well, the reservoir model constructed must conform to the real condition of the oil field, i.e. the well test interpretation model must be the real reservoir model and must consider all kinds of development mechanisms. This model is apparently a very complex flow model and can not be solved using analytical methods. The only way to solve this model is by using a numerical method, and the corresponding well test theory and interpretation method is called numerical well testing. So we can see that the emergence of numerical well testing is necessary.

1.5.1. Basic Concepts of Numerical Well Testing
1.5.1.1. Numerical Well Testing

The so-called numerical well testing refers to the methods using the numerical solution of the well test interpretation model to carry out well test analysis and interpretation. It develops a numerical well test simulator using numerical methods according to the reality of the reservoir. Using this simulator we can simulate exactly the pressure response of the testing well. The solution of the well test interpretation model is the basis for numerical well test interpretation.

The numerical solution of the well test model has many advantages such as:

1. It can solve the well test interpretation model with complex flow mechanisms such as water flooding, miscible drive, gas drive, chemical flooding, thermal drive, and so on.
2. It can deal with boundaries or faults conveniently.
3. It can conveniently deal with complex inner boundary conditions such as variant well bore storage conditions and well bore phase changes and so on.
4. It can consider the impact of production history on the pressure response of the testing well conveniently.
5. It can conveniently consider the effect of multi-well production on the pressure response of the testing well.
6. It can conveniently consider the effect of multi-layer on the pressure response of the testing well.
7. It can conveniently consider the types of the testing well such as production well, injection well, horizontal well, inclined well and various special types of horizontal wells.

At present, the currently used numerical method is the finite difference method.

1.5.1.2. Regional Well Testing

Regional well testing refers to well test interpretation focusing on a region around the testing well or the whole reservoir. The pressure response of the testing well only reflects the regional formation where the pressure around the testing well can reach, so it is not necessary to study the region far from the testing well.

1.5.1.3. Four-Dimensional Well Testing

The four-dimensional well test concept is similar to the four-dimensional seismic concept. That is to say, well test interpretation should consider the testing data at different developmental times, then adjust the reservoir model step by step, and develop a series of reservoir models at different developmental times, and finally form a real four-dimensional reservoir model. It is very important to the oil and gas field development because precise reservoir model is the foundation for all oil and gas field development work. However, the four-dimensional well testing will further enlarge the application range of well test information in reservoir performance description.

If we study well test theory and interpretation methods following this idea of well test analysis, the present well test theory and interpretation methods will be greatly improved.

1.5.2. Steps of Numerical Well Testing Analysis

According to the requirement that the well test theory and interpretation methods must fit the practical oil field condition, the well test interpretation model should be based on a real reservoir model and consider the drive mechanism and production history. Consequently, according to the analysis above, a complete numerical well test should include the following five steps.

1.5.2.1. Building a Complete and Practical Well Testing Interpretation Model

A well test interpretation model includes a reservoir model and a flow mathematical model. According to reservoir geological statistics theory, we can use the results of reservoir fine characterization and well-log multifunctional interpretation information, etc., to build a real reservoir geological model which can consider the heterogeneity of the reservoir. The reservoir model built using the above static information in well test interpretation is an initial reservoir model, which can be adjusted or verified through well test interpretation. The flow mathematical model can be built according to the real conditions in the oil field considering multi-well, multi-phase flow and production history. As to water-flood oil field, the black oil model can be used as the well test interpretation model. For condensate gas reservoirs, in order to consider phase change between fluids, the well test interpretation model must be a compositional model. In terms of a thermal recovery reservoir, the well test interpretation model must be a thermal recovery model which takes into consideration the effect of the temperature of injected thermal fluid. As to a chemical flooded reservoir, the well test interpretation model must be a chemical flooded model. In addition, the consideration of the inner boundary should be identical to the real situation.

1.5.2.2. Solution of the Well Testing Interpretation Model

It is clearly impossible to solve such a complex well test interpretation model using an analytical method. The only way to solve this model is by using a numerical method to develop a corresponding numerical well test simulator to simulate the pressure response of the testing well according to the actual reservoir situation. However, the well test interpretation is an inverse problem which has strict requirements in terms of stability of solution method and computational speed. As a result, when selecting numerical methods to solve the well test interpretation model, we should examine or select the numerical solution methods from these two aspects.

1.5.2.3. Interpretation of the Testing Data

Currently there are many numerical well test interpretation methods just like the available solution methods of the well test interpretation model. However, due to the increase in interpretation parameters and the enlargement of the non-linear degree of the real interpretation model, the requirements for the practical interpretation methods are becoming greater and greater.

1.5.2.4. Verification of the Interpretation Results

Because the effect of production history has been taken into consideration, when verifying the well test interpretation results we can not only verify the dimensional Horner curve and pressure history, but can also verify production data of the testing well such as water cut match and production rate match, etc. This means that we can verify whether the reservoir model determined by well test interpretation coincides with the real production dynamic performance.

1.5.2.5. Interpretation of the Testing Data of Different Wells at Different Development Times

Focusing on the whole reservoir, we can interpret the testing data of different development time to adjust the reservoir model continuously. By doing this, we can develop the corresponding reservoir model at different development times and finally accomplish the idea of building the real 4-D reservoir model.

The numerical well test analysis method involves many different kinds of data and theories, so it can develop a more sophisticated reservoir model and further enlarge the application range of well test interpretation methods.

1.5.3. Technical Route of Numerical Well Testing

According to the above elaboration and assumptions, the technical route diagram of numerical well test analysis method is shown in Fig. 1.5.

1.5.4. Streamline Numerical Well Testing Interpretation Theory and Method

According to the technical route of numerical well test analysis method above, the main research contents are as follows.

1.5.4.1. Building a Reservoir Model

On the basis of reservoir geological description and well-log multifunctional interpretation data, we use the Kriging method to build a reservoir model which includes a static geological model and a production performance model.

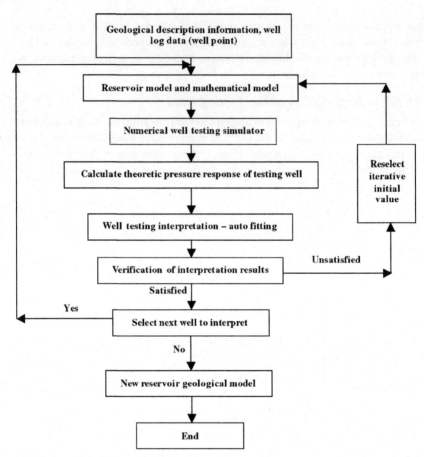

FIGURE 1.5   Technical route of numerical well test analysis method.

1.5.4.2. *Solution Methods of the Well Testing Interpretation Model*

There are many numerical methods to solve flow equations. Among these methods, the finite difference method (which can be classified as rectangular grid, perpendicular bisectors (PEBI) grid and mixed grid according to grid types) and finite element method are most commonly used. However, these methods have two main disadvantages: one is the low computational speed; the other is that numerical dispersion is serious and the stability of computational results is poor. In recent years, the streamline method developed very quickly from the original solution of steady or pseudo-steady flow problems to the solution of transient flow problems. Its characteristics of fast computational speed and good stability of computational results fit just right with the request of well test interpretation for numerical methods. Currently, the streamline method has not been used in well test analysis neither at home nor abroad. The highlights of this book are to develop a suitable well test simulator for a black oil reservoir and a suitable compositional well test simulator for a reservoir with phase change.

1.5.4.3. Well Testing Interpretation Methods

There are many well test interpretation methods including optimization methods (such as the Gauss–Marqudt method which considers constraint condition), genetic algorithms, cybernetics methods, the currently used generalized pulse spectrum method and perturbation methods, and so on. This book adopts the commonly used non-linear regression analysis method.

1.5.4.4. Technical Route of the Streamline Numerical Well Testing Interpretation Method

According to the above technical route of numerical well test interpretation methods, the technical route of the streamline numerical well test interpretation method is proposed (see Fig. 1.6).

1. Building the filtration mathematical model of the production period based on the real reservoir model. The filtration mathematical model of the production period for a water-flooding reservoir is a simplified black oil model (only oil and water phases are considered, the gas phase is not considered); the compositional model is used for a condensate gas and gas injection (CO_2, air, N_2, etc.) reservoir; the chemical flooding model is used for a chemical flooding reservoir; the double-porosity reservoir model is used for a fractured reservoir.
2. Solution of the filtration mathematical model of the production period. The streamline method is used to solve the above filtration mathematical model,

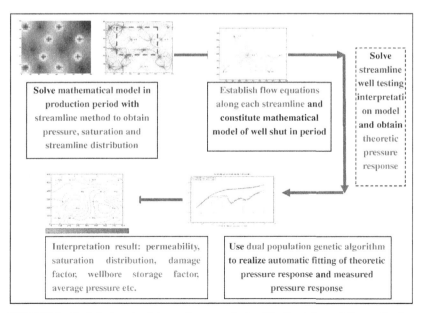

FIGURE 1.6 Technical route of streamline numerical well testing interpretation method.

then fluid saturation distribution, pressure distribution and streamline distribution are obtained as the initial value for the filtration mathematical model of the shut-in period.

3. Building the filtration mathematical model of the shut-in period. The filtration mathematical model is built along each streamline around the testing well, then combine flow equations of all streamlines and take well bore storage, pollution factor and outer boundary condition into consideration to build the filtration mathematical model of the shut-in period. The three-dimensional flow problem is simplified into a one-dimensional flow problem along the streamlines.

4. Solving the filtration mathematical model of the shut-in period. A numerical method is used to solve the mathematical model. Each streamline can have the same or different damage factor (irregularly damaged zone exists in formation).

5. Theoretical pressure response characteristic and parameter sensitivity analysis.

6. Multi-parameter well testing interpretation method.

1.6. CHAPTER SUMMARY

In this chapter, the history of the development of well testing theory is introduced briefly, then the limitations of modern well testing interpretation methods in well testing interpretation models are pointed out; i.e. the effects of geological factors (such as reservoir heterogeneity) and development factors (such as multi-well production, multi-phase fluid flow, production history) could not be taken into consideration in the well test interpretation. The well test interpretation model is very different to the real reservoir model. So the obvious disadvantages of well testing theory and method are the great limitations in their application in the field. Based on further analysis of well testing essence, related concepts of numerical well testing are introduced, and procedures and main research contents of numerical well testing analysis are elaborated.

Chapter 2

Streamline Numerical Simulation Theory and Method

Combined with the streamline method, reservoir numerical simulation can provide a fast and accurate solution for a reservoir model, and in this way streamline numerical simulation theory and method are formed.

2.1. OVERVIEW OF THE STREAMLINE METHOD

In the petroleum industry, the application of the streamline method began in the 1930s. Over several decades of development, both the streamline tracing method and the solution method of the filtration model along the streamline have made great progress. It has evolved from the stream tube method, which can only trace simple two-dimensional cross-section models, to streamline which can trace the real three-dimensional reservoir; from stable streamline to streamline that could be updated to reflect variations of mobility field; from the study of single-phase one-dimensional flow to two-phase three-dimensional flow, and the analytical and numerical solution methods along streamline are developed. In recent years, many scholars have carried out substantial work on the study of the streamline method. Results indicate that the streamline method is suitable for solving reservoir filtration mathematical models and has the characteristics of high calculation speed and good stability.

Streamline is the path of fluid particles flowing from the injection well to the production well. It can be defined as the line which is made of points of equal streamline function values. At any time, the tangential direction of each streamline point is identical to the velocity vector. The geometrical shape of the streamline is changing and it is a one-dimensional flow channel between the injection well and the production well. Flow mechanisms with different displacement processes can be reflected from these channels, and dynamic indexes can be calculated conveniently. Reservoir heterogeneity and displacement characteristics decide different streamline distribution in the regions with different flow velocities. An important assumption to apply to the streamline method is that the flow rate along the streamline is conservative; i.e. the fluid

Streamline Numerical Well Test Interpretation. DOI: 10.1016/B978-0-12-386027-9.00002-2
Copyright © 2011 by Elsevier Ltd

which flows into one end of the streamline must flow out from the other end of the streamline.

While the streamline method is used to solve the reservoir flow problem, three-dimensional or two-dimensional heterogeneity problems could be automatically reduced into a series of one-dimensional, homogeneous, simple problems according to the streamlines in the flow field, then analytical or numerical methods are used to solve these one-dimensional problems along the streamlines. At last, we can gather together all the solutions for streamlines and obtain the whole-reservoir solution.

In reservoir numerical simulation, the essential difference between the streamline method and traditional reservoir numerical simulation is the difference in fluid flow channel. In traditional reservoir numerical simulation, fluid transports among the divided grids in the reservoir. However, in the streamline method, fluid transportation has its own characteristics: it is separated from the basic grid system and moves forward along the streamline, and then the natural flow channel of fluid is formed. Through the transportation of fluid along the continuously changing streamlines, a dynamic adapted grid system is adopted. This dynamic grid system is totally separate from the static grid system divided on the basis of the former reservoir, and the streamline method is used to solve the reservoir filtration model in this dynamic grid system.

Compared with traditional calculation methods, the streamline method has many advantages. In traditional reservoir numerical simulation, the same grids are used to solve pressure and saturation, and fluid only flows along the grid direction. In the streamline method, saturation moves forward along each flow unit of the streamline, and the basis grid blocks are not needed to solve the pressure field of the whole reservoir. This could reduce the influence of the grid division and grid arrangement on the calculation process and results, and lead to the accurate calculation results of streamline method. Meanwhile, the length of time taken to solve the pressure field in the basic grid system reduces significantly, and calculation of saturation along streamlines can take longer, so the calculation speed of the streamline method increases greatly. All these advantages make the streamline method suitable for modern reservoir simulation.

2.2. CALCULATION PROCEDURES OF STREAMLINE NUMERICAL SIMULATOR

Separate the whole-reservoir simulation time into n time phases; the time-step of each phase is Δt^n, then,

$$t^{n+1} = t^n + \Delta t^n$$

Divide the reservoir into grids according to the requirements of numerical simulation, and give initial values of pressure and saturation to each grid according to the initial condition. The procedures of reservoir numerical simulation with the streamline method are shown as follows:

1. In starting time t^n of each new pressure time-step, the implicit pressure explicit saturation (IMPES) method is used to solve the filtration equation which describes multi-phase flow and obtains the pressure field in the grid system.
2. Use the Darcy equation to determine the velocity in each direction of the grid block surface.
3. Use the streamline tracing method to trace streamlines from the injection well to the production well. Calculate the fluid property parameters and the initial saturation distribution of each node in each streamline.
4. Take one-dimensional numerical calculations along each streamline, and obtain the saturation distribution along the streamline at the next time is Δt^{n+1} through moving saturation distribution forward by several time steps Δt_c (one Δt^n). Meanwhile, use the streamline method to solve different dynamic indexes.
5. Average the property parameters of all streamlines in each grid block and obtain the saturation distribution of the grid system at the next time (t^{n+1}).
6. If big events which can change the distribution of streamlines greatly happen (such as well open/shut, well conversion, layer change or drill infill well), streamlines need to be updated and procedures 1~5 above must be repeated. Otherwise, streamline distribution will not change and former streamlines will be used to solve related parameters.

Circulate the above procedures until the calculation end. Collate calculation results and then realize numerical simulation of reservoir production history with the streamline method. The calculation block map of the streamline numerical simulation is shown in Fig. 2.1.

2.3. DISCUSSION OF TIME-STEP

In the process of streamline numerical simulation, the time-step for updating the streamline and solving the pressure field is defined as Δt; the time-step for calculating saturation along the streamline is defined as Δt_c. The values of these two time-steps are different, and so are the determination methods. Δt is determined according to the case whether there are events which can cause streamline change to happen. For the numerical method which explicitly solves the saturation along the streamline, Δt_c is determined by the local stability constraint condition CFL (Courant–Friedrich–Lewy) along the streamline. The mathematical expression of this condition is as below:

$$\Delta t_c \leq \frac{N_c \Delta \tau_{sl}}{v_{max}} \tag{2.3.1}$$

where N_c is the Courant number along the streamline, $N_c \leq 1$; v_{max} is the velocity of the saturation which moves fastest along the streamline, in cm/s; $\Delta \tau_{sl}$ is the difference in node transit time τ along the streamline, s.

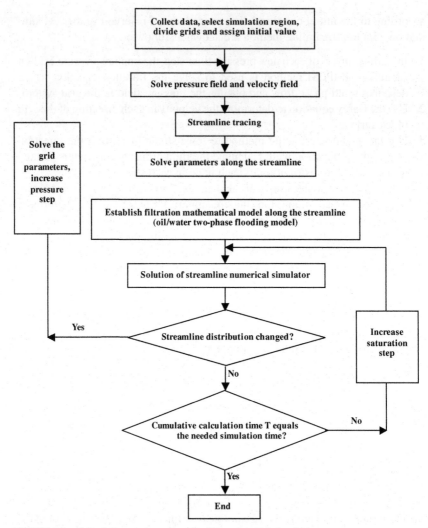

FIGURE 2.1   Calculation process of streamline numerical simulation.

The concept of transit time will be elaborated on later. The physical meaning of the above formula is that the movement distance of the saturation with the highest velocity is less than one τ node in each time-step.

In the traditional reservoir numerical simulation method, the same time-steps are taken to calculate pressure and saturation which are all solved in the basic grid system. If the IMPES method is used for solution, the value of time-step Δt_c depends on the stability constraint condition of the whole CFL, which means time-step is determined by the reservoir grid with the highest flow velocity and this usually makes Δt_c very little. The grids with the highest flow velocities are usually distributed in the region near the well, and the time-steps determined by these grids are obviously too small for the grids which are far away from the well. So the movement distance of the displacement front in each time-step is far less than the optimal grid size, and this obviously reduces the calculation speed. Besides, pressure field is calculated at the same time in each saturation time step Δt_c, which means that $\Delta t = \Delta t_c$. The consumption time for solving the pressure field takes a large percentage of the whole simulation time, so using the IMPES method in traditional numerical simulation method will take a long time. If the implicit method is used, there is no limitation to time-steps, but convergence problems should be considered. So, compared with the explicit method, the calculation speed of the implicit method will not increase significantly. Meanwhile, the numerical divergence extent in the implicit method is more serious and the solution accuracy is worse.

For the streamline method, fluid moves forward along the streamline, so Δt_c and Δt are separated entirely, and commonly $\Delta t >> \Delta t_c$. Meanwhile Δt_c in the streamline method is bigger than the time-steps in the IMPES method. That is, in the solution of the one-dimensional method along the streamline, many small time-steps are used and the solutions move forward with Δt_c along the streamline. In the given simulation period, the solution of several one-dimensional equations along the streamlines is faster than the solution of the whole three-dimensional equation. For each streamline, the one-dimensional solution method along the streamline produces the local constraint condition to time-step and results in the calculation stability along the streamline, which means that the streamline method can be calculated at the fastest speed with the resolution guarantee as a prerequisite.

2.4. STREAMLINE TRACING

Before the streamline method is used to solve the reservoir model, streamlines are generated first in the research sector. According to the definition, the streamline is a line which is made up of points with equal stream function value. Once the stream function values in the region are obtained, streamlines can be produced by connecting the points with equal stream function value. However, this streamline tracing method is only used in the case of simple two-dimensional cross-sections, and it is very difficult to use in real reservoirs.

We do not trace the streamline by solving the stream function in this book. In fact, we use the transit time method proposed by Thiele et al. (1996) to trace the streamline. This method is based on a semi-analytic method which is proposed by Datta-Gupta to simulate the flow of tracer material in heterogeneous and permeable media, and transit time of tracer particles is used to trace streamlines from injection well to production well in grid system. With the known pressure field in the grid system as a premise, the Darcy equation is used to solve the velocity field, and then the streamlines can be traced. This streamline tracing method is semi-analytical because each streamline is made up of a series of streamline segments in single grid, and the streamline segment of each grid is determined by the analytical method.

2.4.1. Calculation of Pressure Field

Since streamline tracing is based on the known pressure and velocity field in the grid system, so the pressure and velocity field should be recalculated after updating the streamlines. The velocity field will be solved after the pressure field is solved implicitly by IMPES method. The method of solving pressure field is elaborated on below.

Fundamental differential flow equations of oil, gas and water phases are described as follows:

$$\nabla \cdot \left(\frac{Kk_{ro}\rho_o}{\mu_o} \left(\nabla p_o - \gamma_{og}\nabla D \right) \right) = \frac{\partial(\phi\rho_o S_o)}{\partial t} \tag{2.4.1}$$

$$\nabla \cdot \left(\frac{Kk_{ro}\rho_{gd}}{\mu_o} \left(\nabla p_o - \gamma_{og}\nabla D \right) + \frac{Kk_{rg}\rho_g}{\mu_g} \left(\nabla p_g - \gamma_g\nabla D \right) \right)$$
$$= \frac{\partial\left[\phi\left(\rho_{gd}S_o + \rho_g S_g\right)\right]}{\partial t} \tag{2.4.2}$$

$$\nabla \left(\frac{Kk_{rw}\rho_w}{\mu_w} \left(\nabla p_w - \gamma_w\nabla D \right) \right) = \frac{\partial(\phi\rho_w S_w)}{\partial t} \tag{2.4.3}$$

where K is the absolute permeability of reservoir porous media; K_{ro}, K_{rg}, K_{rw} are the relative permeabilities of oil, gas and water, respectively; μ_o, μ_g, μ_w are the oil, gas and water viscosities, respectively; ρ_o, ρ_g, ρ_{gd}, ρ_w are the densities of oil, gas, dissolved gas and water, respectively; ϕ is the porosity; S_o, S_g, S_w are the saturations of oil, gas and water, respectively; p_o, p_g, p_w are the pressures of oil, gas and water phases, respectively; D is the depth related to one datum plane; is the Hamilton operator, which takes the gradient of the latter variable; γ_{og}, γ_g, γ_w are the gravities of formation oil, gas and water, respectively (gravity is the product of the density and gravity acceleration).

$$\gamma_{og} = \gamma_o + \gamma_{gd}$$
$$\gamma_o = \rho_o g$$
$$\gamma_{gd} = \rho_{gd} g$$

where γ_o and γ_{gd} are the gravities of surface oil and dissolved gas, respectively, g is the gravity acceleration.

Considering only oil and water phases and the rate term, the fundamental differential flow equations are described as follows:

$$\nabla \cdot \left(\frac{Kk_{ro}\rho_{og}}{\mu_o} \left(\nabla p_o - \gamma_{og}\nabla D \right) \right) + q_o = \frac{\partial(\phi\rho_{og} S_o)}{\partial t} \tag{2.4.4}$$

$$\nabla \cdot \left(\frac{KK_{rw}\rho_w}{\mu_w} \left(\nabla p_w - \gamma_w\nabla D \right) \right) + q_w = \frac{\partial(\phi\rho_w S_w)}{\partial t} \tag{2.4.5}$$

where q_o, q_w are the mass flow rate of injection and production fluid in unit formation volume; ρ_{og} is the oil density.

The density in the oil equation above is $\rho_{og}(\rho_{og} = \rho_o + \rho_{gd})$, which is instead of ρ_o in the former oil equation with consideration of three phases, because the oil density here includes surface oil and dissolved gas.

Apart from the oil/water two-phase fundamental differential flow equations, some auxiliary equations are needed, including:

$$S_o + S_w = 1 \tag{2.4.6}$$

$$p_w = p_o - p_{cow}(S_w) \tag{2.4.7}$$

where p_{cow} is the capillary force between the oil and water phases.

Besides, parameters of the equations are determined with other auxiliary equations. In the case of oil/water two-phase flow, those functions are described as follows:

$$\rho_{og} = \rho_{og}(p_o)$$
$$\rho_w = \rho_w(p_w)$$
$$k_{ro} = k_{ro}(S_w)$$
$$k_{rw} = k_{rw}(S_w)$$
$$\mu_o = \mu_o(p_o)$$
$$\mu_w = \mu_w(p_w)$$
$$p_{cow} = p_{cow}(S_w)$$

The flow and auxiliary equations constitute the simplified oil/water two-phase black oil model.

In the streamline method, the assumptions are described as follows:

1. fluids and rocks are incompressible;
2. diffusion is not considered;
3. capillary pressure and gravity force are ignored.

According to these assumptions, fundamental differential flow Equations 2.4.4 and 2.4.5 are simplified and arranged as follows:

$$\nabla \cdot [\lambda_o \nabla p] + q_o = \phi \rho_{og} \frac{\partial S_o}{\partial t} \tag{2.4.8}$$

$$\nabla \cdot [\lambda_w \nabla p] + q_w = \phi \rho_w \frac{\partial S_w}{\partial t} \tag{2.4.9}$$

where

$$\lambda_o = \frac{K k_{ro} \rho_{og}}{\mu_o} \tag{2.4.10}$$

$$\lambda_w = \frac{K k_{rw} \rho_w}{\mu_w} \tag{2.4.11}$$

For the pseudo three-dimensional model in this research, diverge the model on the plane, then arrange and obtain oil and water phase equations:

$$c_{oi,j} p_{i,j-1} + a_{oi,j} p_{i-1,j} + e_{oi,j} p_{i,j} + b_{oi,j} p_{i+1,j} + d_{oi,j} p_{i,j+1}$$
$$= f_{oi,j} + V_{i,j} \phi \rho_{og} \frac{S_{oi,j}^{n+1} - S_{oi,j}^n}{\Delta t^n} \tag{2.4.12}$$

$$c_{wi,j} p_{i,j-1} + a_{wi,j} p_{i-1,j} + e_{wi,j} p_{i,j} + b_{wi,j} p_{i+1,j} + d_{wi,j} p_{i,j+1}$$
$$= f_{wi,j} + V_{i,j} \phi \rho_w \frac{S_{wi,j}^{n+1} - S_{wi,j}^n}{\Delta t^n} \tag{2.4.13}$$

The coefficients of the oil phase Equation 2.4.12 are shown as follows:

$$c_{oi,j} = T_{oyj-\frac{1}{2}}$$
$$a_{oi,j} = T_{oxi-\frac{1}{2}}$$
$$e_{oi,j} = -(T_{oxi-\frac{1}{2}} + T_{oxi+\frac{1}{2}} + T_{oyj-\frac{1}{2}} + T_{oyj+\frac{1}{2}})$$
$$b_{oi,j} = T_{oxi+\frac{1}{2}}$$
$$d_{oi,j} = T_{oyj+\frac{1}{2}}$$
$$f_{oi,j} = -Q_{oi,j}$$
$$T_{oxi-\frac{1}{2}} = \frac{2\Delta y_j . h \lambda_{oxi-\frac{1}{2}}}{(\Delta x_{i-1} + \Delta x_i)}$$
$$T_{oxi+\frac{1}{2}} = \frac{2\Delta y_j . h \lambda_{oxi+\frac{1}{2}}}{(\Delta x_i + \Delta x_{i+1})}$$
$$T_{oyj-\frac{1}{2}} = \frac{2\Delta x_i . h \lambda_{oyj-\frac{1}{2}}}{(\Delta y_{j-1} + \Delta y_j)}$$
$$T_{oyj+\frac{1}{2}} = \frac{2\Delta x_i . h \lambda_{oyj+\frac{1}{2}}}{(\Delta y_j + \Delta y_{j+1})}$$
$$Q_{oi,j} = V_{i,j} . q_{oi,j} = \Delta x_i . \Delta y_j . h . q_{oi,j}$$

The coefficients of the water phase Equation 2.4.13 are shown as follows:

$$c_{wi,j} = T_{wyj-\frac{1}{2}}$$
$$a_{wi,j} = T_{wxi-\frac{1}{2}}$$
$$e_{wi,j} = -(T_{wxi-\frac{1}{2}} + T_{wxi+\frac{1}{2}} + T_{wyj-\frac{1}{2}} + T_{wyj+\frac{1}{2}})$$
$$b_{wi,j} = T_{wxi+\frac{1}{2}}$$
$$d_{wi,j} = T_{wyj+\frac{1}{2}}$$
$$f_{wi,j} = -Q_{wi,j}$$

$$T_{wxi-\frac{1}{2}} = \frac{2\Delta y_j.h\lambda_{wxi-\frac{1}{2}}}{(\Delta x_{i-1} + \Delta x_i)}$$

$$T_{wxi+\frac{1}{2}} = \frac{2\Delta y_j.h\lambda_{wxi+\frac{1}{2}}}{(\Delta x_i + \Delta x_{i+1})}$$

$$T_{wyj-\frac{1}{2}} = \frac{2\Delta x_i.h\lambda_{wyj-\frac{1}{2}}}{(\Delta y_{j-1} + \Delta y_j)}$$

$$T_{wyj+\frac{1}{2}} = \frac{2\Delta x_i.h\lambda_{wyj+\frac{1}{2}}}{(\Delta y_j + \Delta y_{j+1})}$$

$$Q_{wi,j} = V_{i,j}.q_{wi,j} = \Delta x_i.\Delta y_j.h.q_{wi,j}$$

The pressure field is solved with the implicit method, oil and water phase equations are combined to eliminate the saturation term and obtain an equation only with pressure P; solve this linear algebraic equation group and get the pressure value.

Combine the oil phase equation with the water phase equation and obtain the following formula:

$$c_{i,j}P_{i,j-1} + a_{i,j}P_{i-1,j} + e_{i,j}P_{i,j} + b_{i,j}P_{i+1,j} + d_{i,j}P_{i,j+1} = f_{i,j} \qquad (2.4.14)$$

where $c_{i,j} = Ac_{wi,j} + c_{oi,j}$

$$a_{i,j} = Aa_{wi,j} + a_{oi,j}$$
$$e_{i,j} = Ae_{wi,j} + e_{oi,j}$$
$$b_{i,j} = Ab_{wi,j} + b_{oi,j}$$
$$d_{i,j} = Ad_{wi,j} + d_{oi,j}$$
$$f_{i,j} = Af_{wi,j} + f_{oi,j}$$

where $A = \dfrac{\rho_{og}}{\rho_w}$, which represents oil/water density ratio.

Formula 2.4.4 is the equation for solving the pressure field in the water-flooding model, and the form of the linear algebraic equation group obtained is shown as follows:

$$\begin{bmatrix} & & & d \\ & & b & \\ & e & & \\ a & & & \\ c & & & \end{bmatrix}\begin{bmatrix} P_1 \\ P_2 \\ . \\ . \\ P_N \end{bmatrix} = \begin{bmatrix} f_1 \\ f_2 \\ . \\ . \\ f_N \end{bmatrix} \qquad (2.4.15)$$

The coefficient matrix of the pressure equation is a pentadiagonal sparse matrix which is similar to the form in Equation 2.4.15. There are two methods to solve this kind of linear algebraic equation group, namely the direct method and the iteration method. Since reservoir problems usually have a large scale, the direct method needs a large memory space and the rounding error problem is serious, so the iteration method is used to solve the pressure in the streamline numerical simulator.

From the point of view of practice and easy calculation, the linear successive over-relaxation (LSOR) iteration method is chosen in our study after the comparison. In the relaxation method, the solution is convergent when the relaxation factor satisfies $0 < \omega < 2$ and it is called over-relaxation when $\omega > 1$; i.e. the increment of each point is larger than the required increment for local balance. Otherwise, it is called under-relaxation. The relaxation factor has an optimal value ω_{opt} which can accelerate the convergence rate greatly. Practice proves that this optimal value lies in the range $1 < \omega_{opt} < 2$, and its typical value is 1.25 which belongs to over-relaxation. Usually this kind of method is called the over-relaxation method, or is abbreviated to the relaxation method.

Relaxation methods include point relaxation and line relaxation methods. The point relaxation method starts calculation from one corner and then calculates point by point in a particular order (for example, from left to right, from bottom to top), and the calculation speed of this method is slow. Line relaxation is used for a group of units (a row or a column) and the pressure is obtained at the same time. It is called linear successive over-relaxation because this method is also relaxed to full lines.

Pressure Equation 2.4.4 is used to analyze the process of the LSOR iteration method for calculating pressure field. At first, all grids are ordered in a standard arranged order, and then LSOR calculation is carried out from bottom to top horizontally (as shown in Fig. 2.2). When the $(k + 1)$th iteration reaches the jth row, all pressure values in the jth row are unknown, and the values $p_{i-1,j}^{(k+1)}, p_{i,j}^{(k+1)}, p_{i+1,j}^{(k+1)}$ are waiting for calculation; all pressure values in the $(j - 1)$th row are just calculated, so $p_{i,j-1}^{(k+1)}$ is known; all pressure values in $(j + 1)$th row is $p_{i,j+1}^{(k)}$ of the last iteration step(k), and this value is also known.

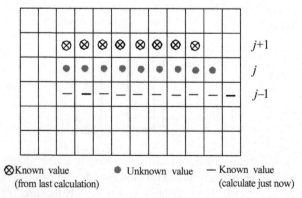

⊗ Known value   ● Unknown value   — Known value
(from last calculation)   (calculate just now)

FIGURE 2.2   LSOR method.

Hence, linear algebraic equations with three unknown variables are obtained when calculating one row. The iteration formula is written as follows:

$$c_{i,j}p_{i,j-1}^{(k+1)} + a_{i,j}p_{i-1,j}^{(k+1)} + e_{i,j}p_{i,j}^{(k+1)} + b_{i,j}p_{i+1,j}^{(k+1)} + d_{i,j}p_{i,j+1}^{(k+1)} = f_{i,j}$$

$$\text{Known} \qquad \text{Unknown} \quad \text{Unknown} \quad \text{Unknown} \quad \text{Known}$$

Move the unknown items to the left and known items to the right, then get

$$a_{i,j}p_{i-1,j}^{(k+1)} + e_{i,j}p_{i,j}^{(k+1)} + b_{i,j}p_{i+1,j}^{(k+1)} = -c_{i,j}P_{i,j+1}^{(k+1)} - d_{i,j}P_{i,j+1}^{(k)} + f_{i,j} = F_{i,j}$$

$$(2.4.16)$$

The coefficient matrix of this formula is a tridiagonal matrix, i.e. the two-dimensional problem is converted into a one-dimensional problem for successive solutions, so the solution problem of the equations is simplified.

After calculating the pressure value $p_{i,j}^*$ of each point in each row, bring in the uniform relaxation factor ω and then obtain the pressure of the $(k+1)$th iteration.

$$p_{i,j}^{(k+1)} = p_{i,j}^{(k)} + \omega(p_{i,j}^{(k+1)} - p_{i,j}^{(k)}) \qquad (2.4.17)$$

Calculate step by step until the convergence rule is achieved.

The above calculation is from bottom to top horizontally; and the successive over-relaxation calculation process from left to right vertically can be derived in the same way.

Grids are arranged in standard order when calculating the pressure field with the LSOR method. The flow chart of the calculation is shown in Fig. 2.3. In the chart, i represents the column number, j represents the row number, k represents iteration times for calculating pressure.

2.4.2. Calculation of Velocity Field

Divide the grid system based on the reservoir simulation region, diverge the flow equations by finite difference method, and use implicit form to solve the pressure field, then the velocity vector of the grid surface can be calculated according to Darcy equation:

$$V_x\left(i \pm \tfrac{1}{2}\right) = -\lambda_{i\pm\frac{1}{2},j,k}[P(i \pm 1,j,k) - P(i,j,k)]/(x_{i\pm1} - x_i) \qquad (2.4.18)$$

$$V_y\left(j \pm \tfrac{1}{2}\right) = -\lambda_{i,j\pm\frac{1}{2},k}[P(i,j \pm 1,k) - P(i,j,k)]/(y_{j\pm1} - y_j) \qquad (2.4.19)$$

$$V_z\left(k \pm \tfrac{1}{2}\right) = -\lambda_{i,j,k\pm\frac{1}{2}}[P(i,j,k \pm 1) - P(i,j,k)]/(z_{k\pm1} - z_k) \qquad (2.4.20)$$

where λ is the mobility sum of each fluid phase, viz $\lambda = \sum_{p=1}^{N_p} \frac{KK_{rp}}{\mu_p}$ (K is the absolute permeability, K_{rp} is the relative permeability of phase p, μ_p is the fluid viscosity of phase p, N_p is the total number of fluids); P is the pressure; x, y, z are

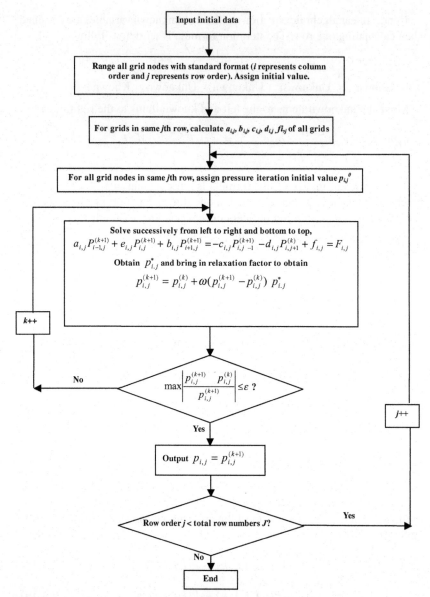

FIGURE 2.3   Flow chart of LSOR method for solving pressure field.

the grid coordinates of grid nodes in three directions, respectively; i, j, k represent the node numbers of three directions x, y, z, respectively.

The velocity of each direction calculated above is the Darcy velocity. In order to trace streamlines, the Darcy velocity should be converted to real velocity. The calculation formula for real velocity is:

$$V_{real} = \frac{V_{Darcy}}{\varphi(i,j,k)} \qquad (2.4.21)$$

where $\phi(i,j,k)$ is the porosity of the grid node.

2.4.3. The Semi-Analytical Method of Streamline Tracing

The basic assumption of the semi-analytical method when tracing streamline from injection well to production well is that the velocity field of each coordinate direction in each block varies linearly and is irrelevant to the velocity of other directions in the same grid. That is, the velocity component of direction x in each grid varies linearly in direction x and is irrelevant to the velocity of directions y and z in the same grid. The same applies for the velocity components of directions y and z. The advantage of this method is that it is analytical and satisfies the material balance equation of the underground reservoir flow field.

For convenience, the semi-analytical method of streamline tracing is described in a two-dimensional system (as shown in Fig. 2.4). Local and origin coordinates are defined in the figure, and assume that the true velocity of grid surface is obtained by the above method.

Velocity V_x of direction x in the grid is

$$V_x = V_{x,o} + m_x(x - x_o) \qquad (2.4.22)$$

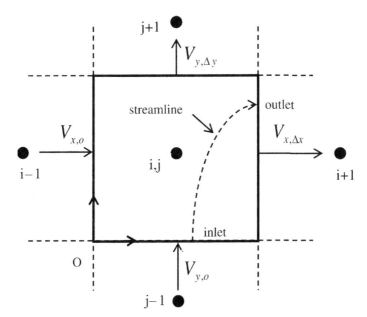

FIGURE 2.4 Schematic of streamline tracing in a two-dimensional system.

where m_x is the velocity gradient of direction x along the grid block, which is calculated as follows:

$$m_x = \frac{v_{x,\Delta x} - v_{x,o}}{\Delta x} \qquad (2.4.23)$$

where Δx is the grid step in direction x.

With the definition of velocity $v_x = dx/dt$, integrate Formula 2.4.22 and determine the time taken by particles to transport from the inlet plane to the outlet plane in direction x:

$$\Delta t_{e,x} = \frac{1}{m_x} \ln\left\{ \frac{V_{x,o} + m_x(x_e - x_o)}{V_{x,o} + m_x(x_{in} - x_o)} \right\} \qquad (2.4.24)$$

where x_{in} is the inlet position; x_e is the outlet position; x_o is the origin position.

If a three-dimensional system is considered, the time taken by particles to arrive at the outlet in directions y and z can be calculated in the same way:

$$\Delta t_{e,y} = \frac{1}{m_y} \ln\left\{ \frac{V_{y,o} + m_y(y_e - y_o)}{V_{y,o} + m_y(y_{in} - y_o)} \right\} \qquad (2.4.25)$$

$$\Delta t_{e,z} = \frac{1}{m_z} \ln\left\{ \frac{V_{z,o} + m_z(z_e - z_o)}{V_{z,o} + m_z(z_{in} - z_o)} \right\} \qquad (2.4.26)$$

Since streamline must flow out from one plane of the grid, so compare the time of the three formulae above and the real outlet position of streamline is determined by the plane with minimum time. Once minimum time is determined, the accurate outlet position of the streamline can be calculated from the transformation of Equations 2.4.24, 2.4.25 and 2.4.26, which are shown as follows:

$$x_e = x_o + \frac{1}{m_x}[V_{x,in} \exp(m_x \Delta t_e) - V_{x,o}] \qquad (2.4.27)$$

$$y_e = y_o + \frac{1}{m_y}[V_{y,in} \exp(m_y \Delta t_e) - V_{y,o}] \qquad (2.4.28)$$

$$z_e = z_o + \frac{1}{m_z}[V_{z,in} \exp(m_z \Delta t_e) - V_{z,o}] \qquad (2.4.29)$$

For the given direction, when grid velocity is constant, $m = 0$. Take Formula 2.4.24 as the example, it can be simplified as

$$\Delta t_{e,x} = (x_e - x_{in})/V_{x,o} \qquad (2.4.30)$$

Formula 2.4.10 can be simplified as

$$x_e = x_o + \Delta t_{e,x} V_{x,o} \qquad (2.4.31)$$

In one grid block, we can predict that the signs of the inlet velocity $V_{x,in}$ and the outlet velocity $V_{x,e}$ in direction x are the same, and the same applies to the situations in directions y and z. Then the case of taking minus logarithm will not occur in Equations 2.4.24, 2.4.25 and 2.4.26.

2.4.4. Time of Flight

Time of Flight (TOF) indicates the time needed when particles move a certain distance s along the streamline. As early as in the 1970s, some authors used the concept of TOF along streamline when they studied the influence of permeability anisotropy on reservoir performance, and they proposed that displacement front is made up of points with the same TOF in different streamlines. In recent years, Datta–Gupta and King (1995) applied TOF into reservoir flow field and gave a more accurate description for TOF.

From the point of view of mathematics, TOF along one streamline τ is defined as

$$\tau(s) = \int_0^s \frac{\phi(\zeta)}{|u_t(\zeta)|} d\zeta \qquad (2.4.32)$$

where ζ is the distance coordinate along the streamline, $\phi(\zeta)$ is the porosity along the streamline, and $u_t(\zeta)$ is the transit velocity along the streamline.

In the semi-analytical method of streamline tracing, the time $\Delta t_{e,i}$ could be calculated when the streamline passes through each grid block and then TOF along the streamline could be estimated. For each grid block i, $\Delta t_{e,i}$ can be calculated through the minimum time which is determined by Formulae 2.4.24, 2.4.25 and 2.4.26, then sum $\Delta t_{e,i}$ of all grids that are passed by the streamline, namely

$$\tau = \sum_{i=1}^{nblocks} \Delta t_{e,i} \qquad (2.4.33)$$

where $\Delta t_{e,i}$ is the TOF of streamline passing grid i; $nblocks$ is the total number of grids passed by the streamline corresponding to the calculated TOF.

TOF is a very important parameter, it can be seen from the latest statement that TOF of streamline passing each grid is needed to solve the diverged differential equations along the streamline. Meanwhile, for the case of streamline variation, saturation still needs to move forward along the new streamline after recalculating the pressure field and retracing the streamline. All streamline properties in the grids should be averaged by TOF of each streamline in order to get fluid property parameters and initial distribution of saturation of the new streamline, then the average property parameters of the grids are obtained. So in the process of streamline tracing, TOF of each streamline node should be recorded.

2.4.5. Flow Rate Distribution of Streamline Tracing

In the reservoir with arbitrary boundary shape and well pattern, the above semi-analytical method is applied to trace streamlines from injection well grid to production well grid. Because in the grid with point source, velocity field can not be seen as piecewise linear, so the streamlines do not start from the center of the injection well grid. Actually, streamlines emit from each surface of the grid in the injection well.

For calculation convenience, the basic principle of rate distribution is to fix the flow rate of each streamline and make streamline number changes with the flow rate of the injection well. For the well with a big injection rate, the number of streamlines that emit from each surface of the block in the injection well will be greater; and for the well with a small injection rate, the number of streamlines that emit from each surface of the grid with injection well will be fewer. In a similar way, more streamlines could be traced in the region with high velocity while fewer streamlines are traced in the region with low velocity.

The rate from the surface of each injection well grid is uniformly distributed and concurs with the velocity field. The rate of each surface is uniform and the rate of each streamline is stable, so streamlines are also uniformly distributed in each plane. The total streamlines emitted from each surface of injection well grid are described as follows:

$$N_{tsl} = \frac{Q_t}{q_{sl}} \qquad (2.4.34)$$

where Q_t is the total rate of injection well; q_{sl} is the rate of each streamline.

2.4.6. Treatment of Grids Without Streamline Passing

In the process of streamline tracing from the injection well to the production well, when the total streamline number is fixed, not all grids in the region will be passed by the streamlines. Because this streamline tracing method determines streamline distribution of several grids in one calculation, most grids have several streamlines passed while some grids have no streamlines passed. However, in order to get grid property parameters by conversion after updating the streamlines, each grid must contain at least one streamline. For calculation convenience, special treatment is needed for those grids without streamline passing.

In the velocity field, take opposite tracing from grids without streamline passing to the injection well and record the coordinate and TOF of each point in opposite tracing, then the property parameters of this kind of grid can be obtained. However, it must be noted that when property parameters in the grid system are solved by conversion, the streamlines obtained by opposite tracing will not influence the properties of other grids with streamline passing. Generally speaking, the flow velocity of grids without streamline passing is very low, so the TOF of these grids is very large. When all kinds of property parameters are converted between streamline unit and grid system repeatedly, these TOF with big values will have a great influence on the property parameters of other grids, and usually unreasonable low saturation values will be assigned to the grids with streamline passing. For this reason, after a TOF is obtained by opposite tracing for grids without streamline passing, when solving the property parameters of grid system by conversion, only the grids without streamlines passing are given new property parameters according to the streamlines obtained by opposite tracing.

During opposite tracing, first trace forward from the center of grids without streamline passing and obtain the next streamline point coordinate in the grid boundary, then take this point as the beginning point of opposite tracing and still use the above semi-analytical method for streamline tracing. It should be noted that the formula of opposite tracing is obtained by converting the streamline tracing formula from the injection well to the production well, then trace until the injection well and record the corresponding streamline node coordinate and TOF. After updating the streamlines, the property parameters of grids without streamline passing can be obtained.

2.5. CALCULATION OF STREAMLINE PARAMETERS

According to the previous streamline tracing algorithm, the shape and distribution of the streamlines, as well as the characteristic parameters along streamlines, can change with time. These streamline parameters mainly include permeability, saturation and so on. Because the streamlines gained by tracing are expressed by a number of streamline nodes, the parameters along the streamlines are parameters of these streamline nodes.

The computational methods of streamline parameters at different streamline node locations are shown as follows.

1. When one streamline point coordinate lies in the grid line of some direction and the other coordinate is in the grid (Fig. 2.5), parameters of streamline notes must be gained according to the property parameters of the two neighboring grids.
2. When the X-axis and the Y-axis of streamline nodes lie in the mesh line at the same time (Fig. 2.6), the streamline parameters of the nodes are

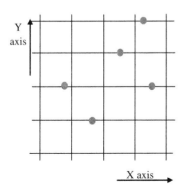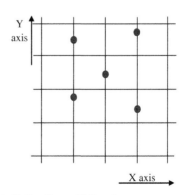

FIGURE 2.5   Streamline nodes of grids paralleling to X direction or Y direction.

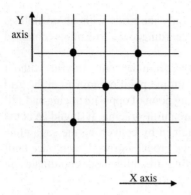

FIGURE 2.6 Streamline nodes lying in nodes of mesh lines.

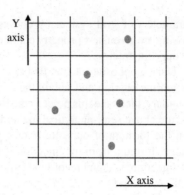

FIGURE 2.7 Streamline nodes lying in grids.

calculated by parameters of the two grids which the streamlines emit from and enter in.

3. When streamline nodes lie in the grids (Fig. 2.7), which is the simplest case, the node parameters of streamlines can be directly expressed by the property parameters of the grids lying at the streamline nodes.

Computational methods are different for different kinds of parameters. Once it is known which grids are needed to calculate the streamline parameters, different parameters can be calculated by the following methods.

1. Absolute permeability. Use the harmonic-mean algorithm, the calculation formula is

$$K_{12} = \frac{2K_1 K_2}{K_1 + K_2} \tag{2.5.1}$$

2. Relative permeability. Adopt the upstream weighting principle. For the point i and its neighboring point i + 1, the upstream weighting principle means that the parameter value equals to the value of point i if fluid flows from point i to point $i+1$, otherwise if fluid flows from point $i+1$ to point i, the parameter value will equal the value at the upstream point $i + 1$. It can be expressed mathematically as follows:

$$k_{rl\,i+\frac{1}{2}} = \begin{cases} k_{rl}(S_{w\,i}) & \text{flow from point } i \text{ to point } i+1 \\ k_{rl}(S_{w\,i+1}) & \text{flow from point } i+1 \text{ to point } i \end{cases}$$

According to this principle, the upstream grid can be determined by the location of the streamline point, and then the relative permeability at the streamline node can take the value of the upstream grid.

3. Saturation. Also adopt the upstream weighting principle. For any parameter related to saturation and pressure in calculation, space values can be determined by the upstream weighting principle.

2.6. STREAMLINE UPDATE

2.6.1. Necessity of Streamline Update

In the process of oil field development, the well pattern and production system are not always fixed. Especially in oil fields with a long production history new stimulated treatments always need to be developed, such as closing producing wells with high water cut, infill well pattern by drilling new wells, re-perforating or modifying layers for old wells, and so on. These treatments can all lead to changes in streamline distribution. If the same streamlines are used in the whole simulation process, these circumstances can not be properly reflected. It is therefore necessary to update streamlines when these events which can greatly change the streamline distribution happen, in order to reflect the displacement dynamics accurately.

2.6.2. Main Procedures of Streamline Update

The main procedures of streamline update are similar to the previous streamline tracing method. The major difference is that the streamline node parameters at present time should be transformed into the grid system before calculating the pressure field, and then parameters of grid system such as pressure and saturation filed can be calculated. The main procedures are shown in Fig. 2.8.

2.7. CALCULATION OF GRID PARAMETERS

In order to calculate along the new streamlines after streamline updating, we must know the initial distribution of property parameters in the new streamlines,

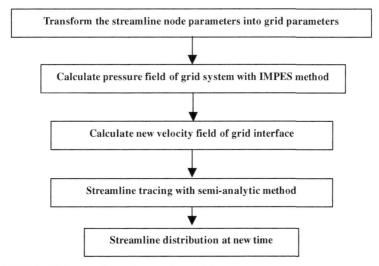

FIGURE 2.8   Main procedures of streamline update.

which is calculated with the property parameters of the grid system. But in the streamline method, the filtration model is solved along the streamlines, so at the time of streamline updating, the old parameter distributions along the old streamlines should be transformed into the grid system, which will act as the known conditions for pressure field recalculation and streamline updating, as well as the continued calculation along the new streamlines.

For most of the grids, there may be more than one streamline, but different property parameters in each grid only have one value. Thus when determining the grid property parameters, the parameters must be calculated using all the streamlines of the grid. The transit time difference of each grid streamline is taken as the weighting factor to calculate the average property parameters of the grid (such as saturation, the total fluidity and the transit time, etc). Taking the grid saturation as an example, the specific calculation method is shown below.

The average saturation of the grid $\bar{S}_{gb}$ can be calculated by

$$\bar{S}_{gb} = \sum_{i=1}^{n_{gb}^{sl}} \omega_i \bar{S}_i^{sl} \tag{2.7.1}$$

where $\bar{S}_i^{sl}$ is the average saturation of the streamline i; n_{gb}^{sl} is the number of the grid streamlines; ω_i is the weighting factor of the streamline i .
ω_i satisfies

$$\sum_{i=1}^{n_{gb}^{sl}} \omega_i = 1 \tag{2.7.2}$$

The weighting factors of each streamline can be calculated by

$$\omega_i = \frac{\Delta\tau_i}{\sum_{i=1}^{n_{gb}^{sl}} \Delta\tau_i} \tag{2.7.3}$$

where $\Delta\tau_i$ is the transit time of streamline i passing through grid:

$$\Delta\tau_i = \tau_{exit} - \tau_{inlet} \tag{2.7.4}$$

where τ_{inlet} and τ_{exit} are the transit time of a streamline at the entry point and the exit point, respectively.

According to the method above, weight each streamline in the grid by their transit time values, and then obtain the property parameters of the grid.

Batycky et al. (1986) have adopted different methods to calculate the weighting factors ω_i, and studied their effects on calculating the property parameters of the grid. The results show that, no matter whether the flow rate through all the streamlines in the same grid are equal or not, and no matter whether each streamline's flow rate is taken into account or not in calculating the weighting factors, the final displacement results calculated by these different weighting factors are of no difference. This means that the choice of the weighting factors does not affect the grid properties and the final solutions are all the same. Therefore, for the reason of simple and practical calculating, Formula 2.7.3 can be chosen to calculate the streamline weighting factors.

On the basis of the formula above, we can obtain the average saturation of grid:

$$\overline{S}_{gb} = \frac{\sum_{i=1}^{n_{gb}^{sl}} \left(\Delta\tau_i \overline{S}_i^{sl} \right)}{\sum_{i=1}^{n_{gb}^{sl}} \Delta\tau_i} \tag{2.7.5}$$

Similarly, the average transit time of the grid can be obtained:

$$\overline{\tau}_{gb} = \frac{\sum_{i=1}^{n_{gb}^{sl}} \left(\Delta\tau_i \overline{\tau}_i \right)}{\sum_{i=1}^{n_{gb}^{sl}} \Delta\tau_i} \tag{2.7.6}$$

where

$$\overline{\tau}_i = \frac{\tau_{inlet} + \tau_{exit}}{2} \tag{2.7.7}$$

As to the grids that only have one streamline passed through, the parameters of the grid can be determined directly according to the property parameters of the streamline.

As mentioned earlier, in the process of tracing streamlines from the injection well to the producing well, some grids may have no streamlines passed through. For such grids, the reverse tracing method toward the injection well can be used to obtain streamlines. In the streamline tracing procedure, we also need to record the transit time at all points of the streamlines. When the streamlines need to be updated, the property parameters of the grids can be calculated using the streamlines obtained by reverse tracing.

2.8. WELL PROCESSING METHOD

In the reservoir numerical simulation, if a grid has a well in it, it does not matter whether the well is a producing well or an injection well; it will be treated as point source or point sink. In streamline numerical simulation, different processing methods will be used in different circumstances. These cases are discussed below.

2.8.1. Well Processing Method in Solving Pressure Field

When solving and updating pressure field, well processing method is the same as the well processing method in solving the pressure field implicitly with the IMPES method in reservoir numerical simulation. The basic thinking is to treat the well as point source or point sink and add a production term to the difference equation of the grid. Therefore, the well processing problem becomes the specific processing problem of the production term in difference equation.

On the other hand, the well processing problem in essence, is to determine the inner boundary condition. So a boundary condition is given to the production well or injection well according to the actual situation, such as fixed bottom-hole pressure or fixed production rate (or injection rate). However, due to the enhancement of non-linear extent in multi-phase flow equation, the non-linear degree of the production term is also increased. Therefore, although the production term is added to the equation as a source point in multi-phase flow condition, its coefficient must be linearly processed for each phase, respectively. In the three-dimensional model, a well always passes through several grids at the same time (as shown in Fig. 2.9). And the surface conditions given are all for the whole well, which means that the production given is the production sum of all the grids passed by the well. But all of the production terms in the difference equations are the production of each specific grid, so the production term processing involves not only the relationship between different phases but also the distribution between different grids.

2.8.1.1. Constant Bottom-Hole Flow Pressure.

Assume the oil output of the grid layer k which passed by well is Q_{ok}, then

$$Q_{ok} = [WI\lambda_o(p - p_{wf})]_k \quad k = 1, 2, ...m \tag{2.8.1}$$

where λ_o is the fluidity of oil; WI is the well index; m is the total number of grids passed by well; p is the grid pressure; p_{wfk} is the pressure of the kth layer grid at the borehole face.

Oil fluidity λ_o is calculated by

$$\lambda_o = \frac{Kk_{ro}\rho_o}{\mu_o} \tag{2.8.2}$$

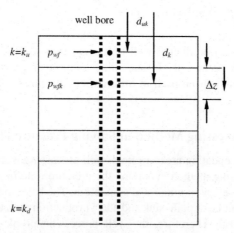

FIGURE 2.9 Sketch map of the well passing through multi-grids.

Well Index (WI) is calculated by

$$WI_k = \frac{2\pi\Delta z_k}{\ln\frac{r_e}{r_w} + S} \tag{2.8.3}$$

where Δz_k is the grid thickness of the kth layer; r_w is the well diameter; S is the skin factor; r_e is the equivalent radius which is calculated by

$$r_e = 0.14\sqrt{\Delta x^2 + \Delta y^2} \tag{2.8.4}$$

Similarly, the production item of the water phase is

$$Q_{wk} = [WI\lambda_w(p - p_{wf})]_k \quad k = 1, 2, \ldots\ldots m \tag{2.8.5}$$

where water phase fluidity λ_w is calculated by

$$\lambda_w = \frac{Kk_{rw}\rho_w}{\mu_w} \tag{2.8.6}$$

After oil and water production are calculated according to the above method, they can be brought into the differential equations as the oil and water production terms, respectively.

2.8.1.2. Constant Liquid Production Rate

If the well is an oil well, the respective oil and water outputs need to be solved first, according to the total liquid output Q_l. Since

$$Q_l = Q_w + Q_o = WI(\lambda_o + \lambda_w)(p - p_{wf}) \tag{2.8.7}$$

Therefore

$$Q_o = \frac{\lambda_o}{\lambda_o + \lambda_w}Q_l \tag{2.8.8}$$

$$Q_w = \frac{\lambda_w}{\lambda_o + \lambda_w}Q_l \tag{2.8.9}$$

The overall oil output of the grids passed by the well is

$$Q_o = \sum_{k=1}^{m} Q_{ok} \tag{2.8.10}$$

Considering Formula 2.8.1, we can get

$$\frac{Q_{ok}}{Q_o} = \frac{[WI\lambda_o(p - p_{wf})]_k}{\sum\limits_{k=1}^{m} [WI\lambda_o(p - p_{wf})]_k} \tag{2.8.11}$$

Therefore, the oil production Q_{ok} of the kth layer grid is

$$Q_{ok} = \frac{[WI\lambda_o(p - p_{wf})]_k}{\sum\limits_{k=1}^{m}[WI\lambda_o(p - p_{wf})]_k} \cdot Q_o \qquad (2.8.12)$$

Similarly, the water production Q_{wk} of the kth layer grid is

$$Q_{wk} = \frac{[WI\lambda_w(p - p_{wf})]_k}{\sum\limits_{k=1}^{m}[WI\lambda_w(p - p_{wf})]_k} \cdot Q_w \qquad (2.8.13)$$

After getting the oil production of all grids, the pressure at the borehole face of the kth layer grid can be solved correspondingly by transforming Formula 2.8.1 into the form below.

$$p_{wfk} = [p - \frac{Q_{ok}}{WI\lambda_o}]_k \qquad (2.8.14)$$

After oil and water production are calculated according to the above method, they can be brought into the oil and water differential equations, respectively, as the production terms.

When solving the production terms above, the formation pressure is explicitly processed, which means that for the grid pressure $p_k = p_{kn}$.

Only the production well processing method is described above. The injection well processing principle and method are basically the same as that of the production well. However, one point should be noted: the pressure at the borehole face of the injection well grid is bigger than the formation pressure of the injection well grid, so the signs in the formulae above should be opposite in application. Another point that should be considered is that when the injection well passes through several grids, production distribution among these grids should be in proportion to the total fluidity of each flowing phase in each grid. Because the different phase fluids originally existing in such grids would be driven away by the injected water, it is most suitable to use the total fluidity coefficient to reflect the flow resistance.

8.2. Well Processing Method in Streamline Tracing

In streamline tracing, for an injection well, first distribute the injection rate into the grids according to the method described in the previous section, then distribute the flow rate of grids into each streamline by the flow distribution method in streamline tracing described in Section 4.4. If the well is closed or abandoned (no injection), no streamlines will emit from the grid surface. If the layer crossed by the well is not completely perforated, there will be also no streamlines emitting from the grid surface of the non-perforated grids. For production well grids, the number of streamlines ending to their surface is determined by streamline tracking. If the production well is closed, abandoned or has non-perforated intervals, it is also assumed that there are no streamlines ending on the grid surface.

8.3. Well Processing Method in Calculation of Grid Parameters

For a grid containing the source or sink term, the velocity field within the grid cannot be regarded as piecewise linear, so the actual streamlines emit from the surface of the injection well grid and enter into the surface of the production well grid. In streamline tracing, the point at the injection well grid surface is the point with the transit time equaling to zero, while the point at the production well grid surface is the last point of the streamline which records the transit time. Therefore, when calculating the property parameters of grids with the source or sink term, the method described in Chapter 2, Section 7, which uses the transit time as the weighting factor for each streamline to calculate the grid properties, is no longer available. We must use other calculation methods.

As mentioned earlier, the weighting factor of each streamline does not affect the calculation of grid property parameters. Therefore, in the calculation of property parameters of a grid containing source or sink term, the reciprocal of streamline number is directly taken as the weighting factor of each streamline, and all streamline property parameters in the grid are directly calculated by the arithmetic average method. The average saturation of the grid $\overline{S}_{gb}$ is calculated as follows.

$$\overline{S}_{gb} = \frac{1}{n_{gb}^{sl}} \sum_{i=1}^{n_{gb}^{sl}} \overline{S}_i^{sl} \tag{2.8.15}$$

where $\overline{S}_i^{sl}$ is the average saturation of streamline i; n_{gb}^{sl} is the number of streamlines in the grid.

In addition to the above circumstances, the source and sink term should also be processed when calculating the saturation along the streamline.

2.9. CHAPTER SUMMARY

This chapter describes the issues related to the streamline numerical simulation method, which include the calculation procedures, time-step selection, the streamline tracing method, streamline parameter calculation as well as the streamline update. Compared with the conventional calculation method currently used, the streamline method has more advantages. The traditional reservoir numerical simulation method uses the same grid to solve the pressure and saturation, and fluids can only flow along the direction of the grid. In the streamline method, the saturation in each flow unit is moved forward along the streamlines and not all basic grids are needed to calculate the pressure field.

So, to a great extent, it reduces the grid effect and ensures the solution accuracy of the streamline method. At the same time, because the times for solving the pressure field in base grid system decrease, and larger time-steps can be used in calculating saturation along the streamlines, so the calculation speed of the streamline method increases greatly.

Streamline Numerical Well Testing Interpretation Model for a Single-Layer Sandstone Waterflooding Reservoir

3.1. BUILDING A STREAMLINE NUMERICAL WELL TESTING INTERPRETATION MODEL FOR A SINGLE-LAYER SANDSTONE WATERFLOODING RESERVOIR

The streamline numerical well testing interpretation model for a single-layer sandstone waterflooding reservoir includes a filtration mathematical model of the production period and a streamline mathematical model of the testing period.

3.1.1. Filtration Mathematical Model of the Production Period

A filtration mathematical model of the production period for a single-layer sandstone waterflooding reservoir is used to simulate the production history and to obtain the pressure and saturation distribution at the time of well testing; the model in this period adopts the simplified black oil model:

1. the fluids in the reservoir are water and oil;
2. rocks and fluids in the reservoir are incompressible;
3. fluid flow in the reservoir obeys the Darcy law;
4. the reservoir is heterogeneous and permeability is isotropic in the plane; and
5. capillary and gravity force are ignored.

Based on the assumptions above, the filtration mathematical model of the production period for a single-layer sandstone waterflooding reservoir is established:

Pressure control equation of oil phase:

$$\nabla \cdot \left(\frac{\rho_o K K_{ro}}{\mu_o} \nabla p \right) + q_o = \frac{\partial (\phi \rho_o S_o)}{\partial t} \tag{3.1.1}$$

Streamline Numerical Well Test Interpretation. DOI: 10.1016/B978-0-12-386027-9.00003-4
Copyright © 2011 by Elsevier Ltd
43

Pressure control equation of water phase:

$$\nabla \cdot \left(\frac{\rho_w K K_{rw}}{\mu_w} \nabla p \right) + q_w = \frac{\partial(\phi \rho_w S_w)}{\partial t} \tag{3.1.2}$$

Saturation normalization equation:

$$S_o + S_w = 1 \tag{3.1.3}$$

Initial condition:

$$\begin{cases} p|_{t=0} = p_i \\ S_w|_{t=0} = S_{wi} \end{cases} \tag{3.1.4}$$

Outer boundary conditions include closed outer boundary and constant pressure outer boundary, which are shown in Equations 3.1.5 and 3.1.6, respectively.

$$\left. \frac{\partial p}{\partial n} \right|_\Gamma = 0 \tag{3.1.5}$$

$$p|_\Gamma = p_e \tag{3.1.6}$$

In the above equations, K is the absolute permeability, μm^2; K_{ro} and K_{rw} are the relative permeabilities of oil and water phases, respectively, dimensionless; P, P_i, and P_e are the formation pressure, initial formation pressure and reservoir outer boundary pressure, respectively (10^{-1}MPa); q_w is the mass water injection rate per unit formation volume and unit time [g/(cm^3•s)]; q_o is the mass oil production rate per unit formation volume and unit time [g/(cm^3•s)]; S_o, S_w, and S_{wi} are oil phase saturation, water phase saturation and initial water saturation, respectively (dimensionless); t is time, s; μ_o and μ_w are the viscosities of oil phase and water phase, respectively (mPa•s); ϕ is the formation porosity, dimensionless; ρ_o and ρ_w are the density of oil and water phase in formation conditions, respectively (g/cm^3).

Inner boundary conditions include constant pressure and constant rate conditions. Here, the constant rate inner boundary condition is mainly considered, which is processed directly by adding the production term to the difference equation of the well grid.

3.1.2. The Streamline Mathematical Model of the Testing Period

The streamline mathematical model of the testing period for a single-layer sandstone waterflooding reservoir is used to simulate the bottom-hole pressure change of the testing well in the testing period and to obtain the theoretical pressure response. The streamline model adopted in this phase is established by grouping the mathematical equations of each streamline around the testing well. Well bore storage effect and skin effect of the testing well, compressibility of rocks and fluids are considered, while the effects of gravity and capillary pressure are ignored.

3.1.2.1. Pressure Control Equation

If j represents the order number of streamline, and the total streamline number emitting from the testing well is N, then the filtration control equation along each streamline is shown as follows:

$$\frac{1}{l_j}\frac{\partial}{\partial l_j}\left(\alpha l_j \frac{\lambda_{tj}}{\phi_j}\frac{\partial p_j}{\partial l_j}\right) = \alpha c_t \frac{\partial p_j}{\partial t} \quad (j=1,2,\cdots,N) \tag{3.1.7}$$

where, l_j is the curve coordinate along the jth streamline, the origin coordinate is the position of the testing well, cm; p_j is the pressure of the jth streamline (10^{-1}MPa); ϕ_j is the reservoir porosity of the jth streamline (dimensionless); λ_{tj} is the reservoir total mobility of the jth streamline [$\mu m^2/(mPa \bullet s)$]; c_t is the total compressibility factor (10^{-1}MPa^{-1}); α is a coefficient, which is the effective thickness for formation with non-uniform thickness and 1 for uniform thickness formation, its meaning in the following chapters will be the same as here.

3.1.2.2. Inner Boundary Condition

The inner boundary condition includes the boundary condition of the production testing well and the boundary condition of the injection testing well, which are expressed, respectively, as follows:

$$\sum_{j=1}^{N}\frac{2\pi}{N}r_w h_1 \cdot \left(\lambda_{t1}\frac{\partial p_j}{\partial l_j}\right)_{l_j=r_w} = -q + C\left[\frac{dp_w}{dt} - Sl_j\frac{d}{dt}\left(\frac{\partial p_j}{\partial l_j}\right)_{l_j=r_w}\right] (j=1,2,\cdots,N)$$

$$\tag{3.1.8}$$

$$\sum_{j=1}^{N}\frac{2\pi}{N}r_w h_1 \cdot \left(\lambda_{t1}\frac{\partial p_j}{\partial l_j}\right)_{l_j=Y_w} = q + C\left[\frac{dp_w}{dt} - Sl_j\frac{d}{dt}\left(\frac{\partial p_j}{\partial l_j}\right)_{l_j=Y_w}\right] (j=1,2,\cdots,N)$$

$$\tag{3.1.9}$$

where, p_w is the well bore pressure of the testing well (10^{-1}MPa); q is the constant production (injection) rate of the testing well in an open well test or the constant production (injection) rate before the shut-in well test (cm^3/s); h_1 is the effective thickness of the testing well point (cm); λ_{t1} is the total mobility [$\mu m^2/(mPa \bullet s)$]; C is the well bore storage coefficient [cm$^3/(10^{-1}$MPa)]; S is the skin factor (dimensionless); r_w is the borehole radius (cm).

3.1.2.3. Outer Boundary Condition

The outer boundary of the streamline is the ending point of the streamline, and it includes reservoir outer boundary and oil/water well outer boundary: reservoir outer boundary is processed in the same way with the traditional well testing, while oil/water well outer boundary condition is determined by the working system of the well where the streamline reaches.

3.1.2.4. Initial Condition

Pressure, saturation and streamline distribution of the testing period can be obtained by solving the filtration mathematical model of the production period, which are taken as the initial condition of the streamline mathematical model of the testing period. The resolution of the filtration mathematical model of the production period is elaborated upon in Chapter 2.

3.1.2.5. Saturation Equation

During the testing period, due to the influence of compressibility of formation and fluids and the neighbor wells, the fluids in the flow zone which are controlled by the testing well still flow, and the mobility distribution in formation changes constantly, so the saturation parameter of the streamline nodes in the above equations should be updated continually. Based on the water phase saturation, the relevant parameters of each node of each streamline in the well testing interpretation model are calculated according to the water saturation; the equation used is

$$\frac{\partial S_w}{\partial t} + \frac{\partial f_w}{\partial \tau} = 0 \qquad (3.1.10)$$

where, τ is transit time (s); f_w is water cut (dimensionless).

The derivation of Equation 3.1.10 and the meaning of τ will be elaborated upon in the following chapter.

3.2. SOLUTION OF THE STREAMLINE NUMERICAL WELL TESTING INTERPRETATION MODEL FOR A SINGLE-LAYER SANDSTONE WATERFLOODING RESERVOIR

3.2.1. Solving Method for the Filtration Mathematical Model of the Production Period

The streamline method is used to solve the saturation of filtration mathematical model for a single-layer waterflooding reservoir in the production period. This method can reduce the complex two-dimensional or three-dimensional problem into a one-dimensional problem along the streamline, then solve the one-dimensional problem along the streamline and obtain the solution in the whole research zone by integrating the solutions of all streamlines.

In recent years, a large amount of work on the streamline method has been carried out by many researchers, which indicates that the streamline method is a quick and effective method and is very suitable for reservoir engineering calculations. Compared with the traditional calculation methods, using the streamline method to solve the reservoir production history has many advantages.

In traditional reservoir numerical simulation calculation methods, the same grids are used to solve pressure and saturation, and fluids only flow along the grid direction; while in the streamline method, the saturation of each flow unit will be moved forward along the streamline and there is no need to use all the basic grid blocks for calculation. In this way, the influence of grid division and arrangement on the calculation process and results decreases greatly, and the calculation results of the streamline method are more accurate. Meanwhile, since the time taken for solving the pressure field in a basic grid system is obviously reduced, and large time-steps can be used to calculate the saturation along the streamline, the calculation speed of the streamline method is faster than traditional numerical simulation methods.

3.2.1.1. Derivation of the Saturation Equation Along Streamline

If the streamline method is used to solve the filtration mathematical model of the production period, firstly the two-dimensional or three-dimensional model based on the grid system should be converted to a one-dimensional model along the streamline. The conversion method is described as follows:

Expand the right terms in Equations 3.1.1 and 3.1.2:

$$\frac{\partial(\phi\rho_o S_o)}{\partial t} = \rho_o S_o \frac{\partial \phi}{\partial t} + \phi S_o \frac{\partial \rho_o}{\partial t} + \phi\rho_o \frac{\partial S_o}{\partial t} \tag{3.2.1}$$

$$\frac{\partial(\phi\rho_w S_w)}{\partial t} = \rho_w S_w \frac{\partial \phi}{\partial t} + \phi S_w \frac{\partial \rho_w}{\partial t} + \phi\rho_w \frac{\partial S_w}{\partial t} \tag{3.2.2}$$

While the compressibility of fluids and rocks are not considered:

$$\frac{\partial \rho_o}{\partial t} = 0$$

$$\frac{\partial \rho_w}{\partial t} = 0 \tag{3.2.3}$$

$$\frac{\partial \phi}{\partial t} = 0$$

Substitute Equation 3.2.3 into Equations 3.2.1 and 3.2.2:

$$\frac{\partial(\phi\rho_o S_o)}{\partial t} = \phi\rho_o \frac{\partial S_o}{\partial t} \tag{3.2.4}$$

$$\frac{\partial(\phi\rho_w S_w)}{\partial t} = \phi\rho_w \frac{\partial S_w}{\partial t} \tag{3.2.5}$$

Then, Equations 3.1.1 and 3.1.2 can be converted to

$$\nabla \cdot \left[\frac{KK_{ro}\rho_o}{\mu_o} \nabla p \right] + q_o = \phi\rho_o \frac{\partial S_o}{\partial t} \tag{3.2.6}$$

$$\nabla \cdot \left[\frac{KK_{rw}\rho_w}{\mu_w} \nabla p \right] + q_w = \phi\rho_w \frac{\partial S_w}{\partial t} \tag{3.2.7}$$

When l represents the phase subscript, and other source/sink terms do not exist along the streamline except the well node, then the two equations above could be written together as

$$\nabla \cdot \left(\frac{KK_{rl}}{\mu_l} \nabla p \right) = \phi \frac{\partial S_l}{\partial t} \qquad (3.2.8)$$

The Darcy equation is used to obtain the flow velocity of the oil phase (u_o), water phase (u_w) and the total flow velocity (u_t), which are described as follows:

$$u_o = -\lambda_o \nabla p \qquad (3.2.9)$$
$$u_w = -\lambda_w \nabla p \qquad (3.2.10)$$
$$u_t = -\lambda_t \nabla p \qquad (3.2.11)$$

where

$$\lambda_o = \frac{KK_{ro}}{\mu_o} \qquad (3.2.12)$$

$$\lambda_w = \frac{KK_{rw}}{\mu_w} \qquad (3.2.13)$$

$$\lambda_t = \lambda_w + \lambda_o = \frac{KK_{rw}}{\mu_w} + \frac{KK_{ro}}{\mu_o} \qquad (3.2.14)$$

Saturation calculation is based on the water phase equation in the study, and the oil saturation is calculated according to the water saturation. Substitute phase subscript l with w in Equation 3.2.8 then the water phase equation is obtained as follows:

$$\nabla \cdot (\lambda_w \nabla p) = \phi \frac{\partial S_w}{\partial t} \qquad (3.2.15)$$

Equation 3.2.15 can be converted to

$$\nabla \cdot \left[-\left(-\lambda_t \cdot \frac{\lambda_w}{\lambda_t} \nabla p \right) \right] = \phi \frac{\partial S_w}{\partial t} \qquad (3.2.16)$$

With the definition of water cut

$$f_w = \frac{\lambda_w}{\lambda_w + \lambda_o} = \frac{\lambda_w}{\lambda_t} \qquad (3.2.17)$$

Equation 3.2.16 is further converted to

$$\nabla \cdot (-u_t \cdot f_w) = \phi \frac{\partial S_w}{\partial t} \qquad (3.2.18)$$

Expand the left term of Equation 3.2.18:

$$-u_t . \nabla(f_w) + f_w . \nabla(-u_t) = \phi \frac{\partial S_w}{\partial t} \qquad (3.2.19)$$

Because the compressibility of fluids and rocks are not considered along the streamline (passive field), so

$$\nabla \cdot (-u_t) = 0 \tag{3.2.20}$$

Then Equation 3.2.19 can be converted to

$$-u_t \cdot \nabla (f_w) = \phi \frac{\partial S_w}{\partial t} \tag{3.2.21}$$

The water phase flow equation of the one-dimensional coordinate (ζ) along the streamline can be written according to Equation 3.2.21:

$$(-u_t) \cdot \frac{\partial f_w}{\partial \zeta} = \phi \frac{\partial S_w}{\partial t} \tag{3.2.22}$$

Define the transit time along the streamline as

$$\tau(s) = \int_0^s \frac{\phi(\zeta)}{|u_t(\zeta)|} d\zeta \tag{3.2.23}$$

Differentiate the two sides for s in Equation 3.2.23

$$\frac{\partial \tau}{\partial s} = \frac{\phi}{|u_t|} \tag{3.2.24}$$

Equation 3.2.24 can be converted to

$$-u_t = -\phi \frac{\partial s}{\partial \tau} \tag{3.2.25}$$

Substitute Equation 3.2.25 into Equation 3.2.21:

$$-\phi \frac{\partial s}{\partial \tau} \cdot \frac{\partial f_w}{\partial \zeta} = \phi \frac{\partial S_w}{\partial t} \tag{3.2.26}$$

Because the coordinate s is equivalent to ζ along the streamline, remove the equivalent terms in numerator and denominator and obtain

$$\frac{\partial S_w}{\partial t} + \frac{\partial f_w}{\partial \tau} = 0 \tag{3.2.27}$$

Equation 3.2.27 is the water phase saturation equation of the oil/water two-phase streamline model.

Initial boundary condition:

$$S_w(s, 0) = S_{wi} \quad s \geq 0 \tag{3.2.28}$$

Inner boundary condition:

$$S_w(0, t) = S_{w,inj} \quad t > 0 \tag{3.2.29}$$

where s is the curve coordinate along the streamline, which is one-to-one correspondent with τ in each streamline and the origin coordinate is the starting point of streamline (cm); S_{wi} is the initial water saturation of each node in streamline (fraction); $S_{w,inj}$ is the water saturation at the injection well (fraction).

Equations 3.2.27 to 3.2.29 constitute the one-dimensional streamline model of saturation resolution for a single-layer sandstone waterflooding reservoir. The unknown variable in this model is water saturation S_w, and the number of equation is one, so the equations are closed and the model can be solved.

3.2.1.2. Differential Solution of the Saturation Equation

The IMPES method is used to solve the pressure equation in the filtration mathematical model of the production period for a single-layer sandstone waterflooding reservoir. The numerical method is used to solve the saturation equation, replacing the derivative by finite difference in space, then Equation 3.2.27 becomes:

$$\frac{S_{wi}^{n+1} - S_{wi}^n}{\Delta t} + \frac{f_{wi+\frac{1}{2}} - f_{wi-\frac{1}{2}}}{\Delta \tau} = 0 \qquad (3.2.30)$$

Move the unknown variables to the left and the known variables to the right:

$$S_{wi}^{n+1} = S_{wi}^n - \frac{\Delta t_{sl}^{n+1}}{\Delta \tau_{sl}} \left(f_{wi+\frac{1}{2}}^n - f_{wi-\frac{1}{2}}^n \right) \qquad (3.2.31)$$

where S_{wi}^{n+1} and S_{wi}^n are the water saturation of streamline node i at time $n+1$ and n, respectively (fraction); $f_{wi\pm\frac{1}{2}}^n$ is the water cut of node $i \pm \frac{1}{2}$ at time n (fraction); Δt_{sl}^{n+1} is the current time-step along the streamline (s); $\Delta \tau_{sl}$ is the difference in flight time between the streamline nodes (s).

Because central difference is used for water cut, the difference of flight time should be between node $i \pm \frac{1}{2}$ and node $i - \frac{1}{2}$. However, the flight time recorded in streamline tracing is the flight time at each streamline node, so the difference of flight time at the half node is calculated as follows:

$$\Delta \tau_{sl} = \left(\tau_{i+\frac{1}{2}} - \tau_{i-\frac{1}{2}} \right) = \frac{1}{2} (\Delta \tau_i - \Delta \tau_{i-1}) \frac{1}{2} (\tau_{i+1} - \tau_i) - \frac{1}{2} (\tau_i - \tau_{i-1})$$

$$= \frac{1}{2} (\tau_{i+1} + \tau_{i-1}) - \tau_i$$

$$(3.2.32)$$

3.2.1.3. Streamline Update

In the whole production history, well pattern and production system vary constantly. Especially for the oil fields with a long production time, new stimulation and potential tapping treatments are often carried out, such as shutting in high water cut wells, arranging infill wells, reperforating the old wells or converting production layers and so on. These treatments will change the distribution of streamline and so the streamline should be updated.

3.2.2. Solving Method of the Streamline Mathematical Model of the Testing Period

The difference method is used to solve the streamline mathematical model of the testing period. Discrete inner boundary condition Equation 3.1.8 or 3.1.9 and get difference Equation 3.2.33, discrete pressure control Equation 3.1.7 and obtain difference Equation 3.2.34, discrete outer boundary condition Equation 3.1.10 and get Equation 3.2.35. Then the difference equations of each streamline can be obtained as follows:

$$a_{ji}p_{ji}^{n+1} + c_{ji}p_{ji+1}^{n+1} = g_{ji} \cdot \ (i = 1) \tag{3.2.33}$$

$$b_{ji}p_{ji-1}^{n+1} + a_{ji}p_{ji}^{n+1} + c_{ji}p_{ji+1}^{n+1} = g_{ji} \cdot \ (i = 2, \cdots, N_j) \tag{3.2.34}$$

$$b_{ji}p_{ji-1}^{n+1} + a_{ji}p_{ji}^{n+1} = g_{ji} \cdot \ (i = N_j) \tag{3.2.35}$$

where, a_{ji}, b_{ji}, c_{ji}, g_{ji} are dispersion coefficients.

Get the difference equations of all the streamlines together and obtain the linear equation group:

$$AP = B \tag{3.2.36}$$

In the formula above, if there are N streamlines emitting from the testing well and each streamline has N_J nodes, then P is the pressure solution vector with $\sum_{j=1}^{N} N_j$ nodes; B is the constant term of the linear equation group; A is the coefficient matrix.

A is a sparse matrix with the size of $\sum_{j=1}^{N} N_j \times \sum_{j=1}^{N} N_j$. The diagram below shows the coefficient matrix structure of the difference equations in the streamline mathematical model for a single-layer reservoir, in which the little-partitioned matrix is the coefficient matrix of the difference equations for the first streamline in the mathematical model (Fig. 3.1).

The iterative or chasing method can be used to solve the linear Equations (3.2.36).

It should be pointed out that the mobility parameter of streamline nodes in the testing period changes constantly, so this parameter should be updated at each time-step. This can be realized by solving saturation Equation 3.2.27, and the solving method is the same as that in the production period.

Through solving the streamline mathematical model of the testing period for a single-layer waterflooding reservoir, we can get the theoretical (calculated) pressure response of the testing well. Then the well testing interpretation

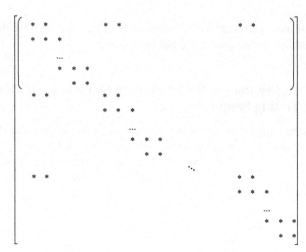

FIGURE 3.1   Coefficient matrix structure of the difference equations in the streamline mathematical model for a single-layer reservoir.

parameters in the influence area of the testing well can be obtained by matching the real pressure data with the calculated pressure response.

3.3. CALCULATION METHOD OF THE STREAMLINE NUMERICAL WELL TESTING INTERPRETATION MODEL FOR A SINGLE-LAYER SANDSTONE WATERFLOODING RESERVOIR

The building of a numerical well testing interpretation model for a single-layer sandstone waterflooding reservoir includes two phases: one is the model building for the production period and the other is the model building for the testing period. The solving process includes the following procedures:

1. Data preparation. Choose the research subject and collect the field data, including geological static data, development dynamic data and pressure-measuring data of the testing well, etc.
2. Reservoir modeling. Discrete the reservoir parameters with the initial parameter value of wells and build the geological model of the target reservoir.
3. Establish the filtration mathematical model of the production period.
4. Use the IMPES method to solve the pressure control equation of the above model and obtain the pressure and saturation distribution under the special working system.
5. Take streamline tracing based on the pressure and saturation distribution obtained above, and then calculate the streamline node parameters.
6. Establish the saturation equation along the streamline; solve the equation by the difference method to obtain the saturation distribution at the new time.
7. Judge whether an event like a change in the working system, which can lead to a change in streamline distribution, takes place. If this kind of event

happens, convert the saturation parameter of the streamline nodes into each grid and recalculate pressure distribution, trace streamlines and calculate the streamline node parameters.

8. Calculate until the testing time with the streamline method and get the pressure and saturation distribution at the time of well testing.
9. Modify the initial geological model of the production period by matching the water cut of the single well and the total region.
10. When the fitting of production indications (water cut, pressure) satisfies the accuracy requirement, calculate and output the correct pressure and saturation distribution at the time of well testing.
11. Establish the streamline mathematical model of the testing period along all streamlines from the testing well, and use the pressure and saturation distribution at the end of the production period to calculate and obtain the pressure and saturation of each streamline node.

Figure 3.2 shows the calculation flow chart of solving the streamline numerical well testing interpretation model for a single-layer waterflooding reservoir.

3.4. CORRECTNESS VERIFICATION OF THE STREAMLINE NUMERICAL WELL TESTING INTERPRETATION MODEL FOR A SINGLE-LAYER SANDSTONE WATERFLOODING RESERVOIR

3.4.1. Drawdown Test Simulation and Results Analysis in a Homogeneous Infinite Reservoir

In order to verify the validity of the above solution method, take a five-spot homogeneous reservoir as the calculation example, degenerate two-phase flow into single-phase flow, use very big well span to simulate the infinite reservoir and then simulate the drawdown test. Figure 3.3 shows the dimensional pressure and pressure derivative bilogarithmic diagram when well bore storage factor and skin factor all equal to zero. Figure 3.4 shows a dimensional pressure and derivative bilogarithmic diagram while well bore storage is $2\,m^3\,/\,MPa$ and skin factor is zero. The points in the figure represent the calculation results of this method, and the lines represent results calculated from the analytical solution. It can be seen that the calculation results of the two methods are very consistent, which proves that the streamline numerical well testing interpretation model is correct for a single-layer homogeneous infinite reservoir (single phase flow).

3.4.2. Drawdown Test Simulation and Results Analysis in Homogeneous Waterflooding Reservoir

The results above have proven the validity of the streamline numerical well testing interpretation model with single-phase flow in a single-layer homogeneous infinite reservoir, and this part will verify the validity of the streamline numerical well

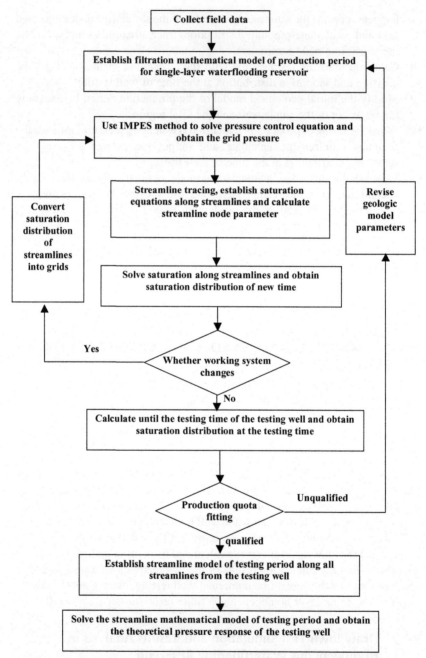

FIGURE 3.2 Calculation chart of streamline numerical well testing theoretical pressure response for a single-layer sandstone waterflooding reservoir.

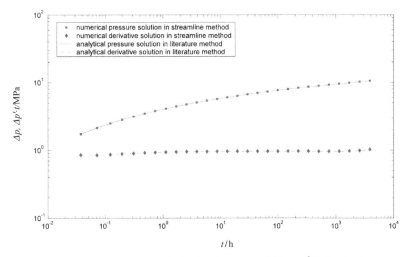

FIGURE 3.3   Pressure and pressure derivative diagram with $C = 0\,m^3/MPa$, $S = 0$.

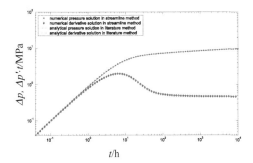

FIGURE 3.4   Pressure and pressure derivation diagram with $C = 2\,m^3/MPa$, $S = 0$.

testing interpretation model with two-phase flow. Establish a typical model of an inverted nine-spot pattern in a homogeneous reservoir, liquid production rate of oil well P1~P8 is $50\,m^3/d$, liquid injection rate of water well I is $400\,m^3/d$. The oil/water relative permeability curve is shown in Fig. 3.5.

In the same production condition, take a different oil/water viscosity ratio (fix water viscosity and change oil viscosity), shut in the injection well I after 180 days waterflooding and simulate the drawdown test. The calculation results are shown in Figs 3.6–3.9.

The pressure response curves with oil/water viscosity ratios taking 10, 20,100 and 200, respectively, are shown in Fig. 3.10. It can be seen that, during the middle part of each curve, the slope of the pressure derivative has a non-zero upwarp tendency. This upwarp phenomenon is caused by the decrease of the total fluid mobility which can be seen from the distribution of saturation in the grids or streamlines. The bottom pressure response curve in the diagram is

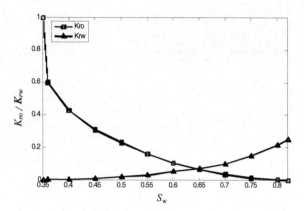

FIGURE 3.5   Oil/water relative permeability curve.

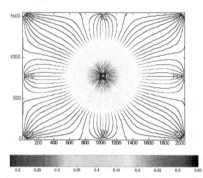

FIGURE 3.6 Oil saturation distribution when oil/water viscosity ratio is 20.

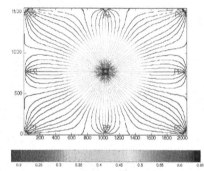

FIGURE 3.7 Oil saturation distribution when oil/water viscosity ratio is 10.

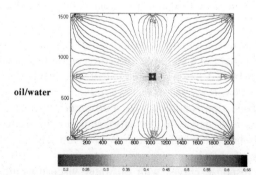

FIGURE 3.8   Oil saturation distribution when oil/water viscosity ratio is 200.

obtained when the oil/water viscosity ratio is the smallest (10), compared with the other three situations, physical properties of oil and water are closest, so the slope of the pressure derivative upwarp in the middle part is the smallest. It can

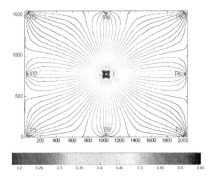

FIGURE 3.9   Oil saturation distribution when oil/water viscosity ratio is 100.

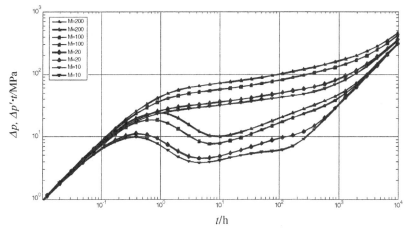

FIGURE 3.10   Pressure response with different oil/water viscosity ratio M.

also be seen from the figure that, with the increase in oil viscosity (water viscosity is fixed), the total mobility of oil and water phase decreases, the pressure response curve rises entirely, and the occurrence time of the radial flow straight segment in the middle part of pressure response delays.

The above analysis demonstrates that the results of the streamline numerical well testing interpretation model fit the actual situation of oil/water two-phase flow, and the validity of the model for two-phase flow is proved.

3.4.3. Drawdown Test Simulation and Results Analysis in Heterogeneous Reservoir

In order to further verify the validity of the streamline numerical well testing interpretation model in heterogeneous conditions, a typical model of an inverted five-spot pattern is established. Permeability and porosity in the model are heterogeneously distributed (see Figs 3.11 and 3.12). The liquid

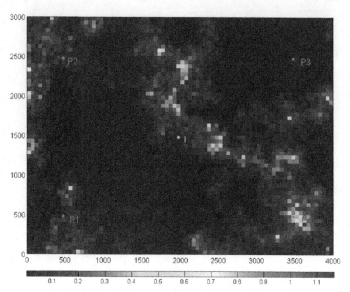

FIGURE 3.11   Permeability distribution /(μm²).

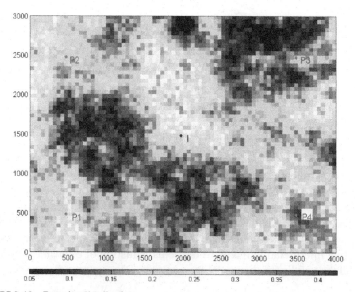

FIGURE 3.12   Porosity distribution.

production rate of oil well P1~P4 is $80\,m^3/d$, liquid injection rate of water well I is $300\,m^3/d$. Oil and water phases use the same physical property parameters and phase permeability, then the model is degenerated into a single-phase flow model, which is convenient for the independent study of the formation parameters.

In this model, water injection well I is taken as the testing well, shut in this well after a certain period of production, and simulate the pressure drawdown. The calculation results are shown in Figs 3.13 and 3.14.

It can be seen from Figs 3.15 and 3.16 that streamlines concentrate in the area of large permeability and porosity, which demonstrates that the distribution pattern of streamlines can reflect the variation characteristics of the formation parameters, and the streamline numerical well testing interpretation model can consider the variation of formation parameters such as permeability etc.

It can be seen from Fig. 3.17, before pressure response achieves outer boundary (neighbor wells), the pressure derivative curve does not display one

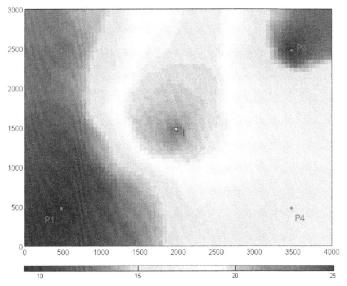

FIGURE 3.13 Pressure distribution at the time of the testing well shut-in /(MPa).

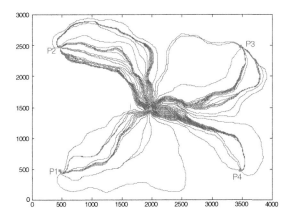

FIGURE 3.14 Streamline distribution at the time of the testing well shut-in.

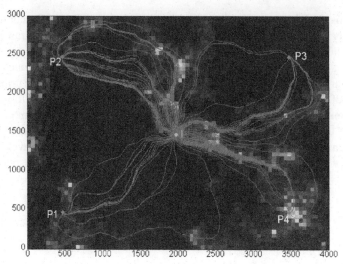

FIGURE 3.15   Stacking chart of streamline and permeability field.

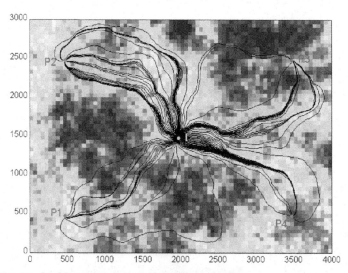

FIGURE 3.16   Stacking chart of streamline and porosity field.

straight segment but an upwarp tendency with different slopes, which is caused by the heterogeneity of formation parameters (permeability and porosity, seen in Figs 3.15 and 3.16) in the reservoir where the streamlines pass through. It can be seen from the building and solving process of the streamline numerical well testing interpretation model that heterogeneous parameters of formation can be converted into node parameters of the model, and the pressure response curves obtained from this model can reflect the variation of these parameters, which illustrates that the streamline numerical well testing interpretation model can

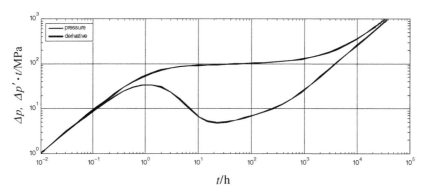

FIGURE 3.17   Theoretical pressure response of the testing well I.

consider the effect of formation heterogeneity and it has a better sensitivity to the formation heterogeneous parameters.

It can be shown from the distribution diagram of permeability and porosity that the main reason for the upwarp tendency in the pressure response derivative curve of the testing well is the change of formation parameters where the streamlines pass through. However, each streamline flows through many grids and it is difficult to judge the change in which grids cause the tendency. Based on the concepts of pressure diffusion and investigation radius created by instantaneous source/sink, the position of the pressure response of the testing well with the maximum value can be calculated at any time. Investigation radius (or pressure front) is defined as: apply a pressure pulse to the testing well at a certain time-point, the position of the pressure pulse when the pulse response achieves the maximum value.

For a homogeneous reservoir, the calculation formula of investigation radius is written as

$$r_i = 2\sqrt{\frac{Kt}{\phi \mu c_t}} \tag{3.4.1}$$

where, r_i is the investigation radius of homogeneous reservoir at any time, cm.

For a heterogeneous reservoir, the investigation radius satisfies the following formula:

$$\int_0^{l_i} \frac{ds}{\sqrt{\frac{K(x)}{\phi(x)\mu c_t}}} = 2\sqrt{t} \tag{3.4.2}$$

where l_i is the investigation radius of heterogeneous reservoir at any time, cm.

Take four streamlines in this model as examples: the pressure front position (point 1 to point 4 in Fig. 3.18) of each streamline could be obtained using

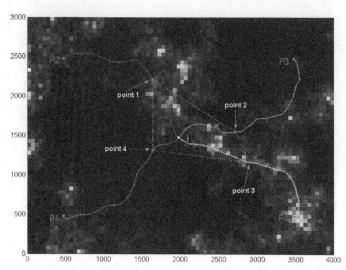

FIGURE 3.18   Pressure front of each streamline at a certain time after the testing well shut-in.

Formula 3.4.2 at any time after the testing well shut-in, which can be used to analyze and verify the variation causes of the pressure response curve pattern; the dashed line in Fig. 3.18 shows the pressure front at a certain time.

With the above method, we can not only get the exact location of the pressure front in a heterogeneous reservoir at any time, but we can also obtain the required time when the pressure front achieves any point of the streamline. As seen in Fig. 3.19, take one point in each streamline as the observation point (point 1 to point 4), use Formula 3.4.2 to calculate the time when pressure front arrives at

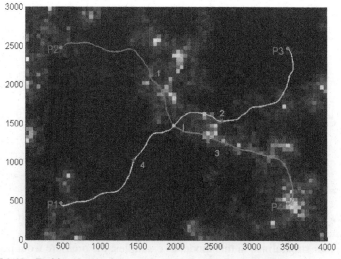

FIGURE 3.19   Position sketch of the random points taken in streamlines.

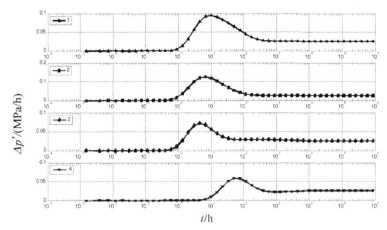

FIGURE 3.20 Pressure derivative curves of each observation point in streamlines.

these four points (which are 5.2 h, 3.8 h, 4.1 h and 40.1 h, respectively), and they are consistent with the variation characteristics of the pressure derivative curves of the four points. With a similar method, the time of pressure front achieving neighbor well P1, P2, P3 and P4 can be solved (232.5 h, 122.1 h, 97.0 h and 25.8 h, respectively), which is significant to the interference test (Fig. 3.20).

3.5. PRESSURE RESPONSE CHARACTERISTICS OF THE STREAMLINE NUMERICAL WELL TESTING INTERPRETATION MODEL FOR A SINGLE-LAYER SANDSTONE WATERFLOODING RESERVOIR

3.5.1. Influencing Characteristics of Damage Factor on Pressure Response

The following shows the influence of different damage factors on well testing pressure response during single-phase flow conditions, in which the five-spot pattern is used as the calculation example. Basic parameters are shown in Table 3.1. Figure 3.21 demonstrates the pressure and derivative curves with skin factors taking −3, 0, 5 and 10, respectively.

TABLE 3.1 Basic Parameters

Reservoir area (m)	Oil formation thickness (m)	Porosity (%)	Permeability (μm^2)	Production rate (m^3/d)	Initial formation pressure (MPa)
1890 × 1890	10	30	800	42	19.2

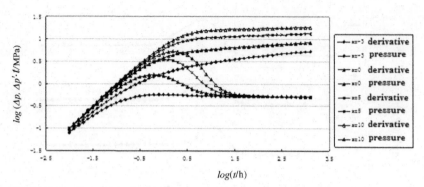

FIGURE 3.21   Pressure and derivative curves with different damage factors.

3.5.2. Influencing Characteristics of Well Bore Storage Factors on Pressure Response

The following demonstrates the influence of different well bore storage factors on well testing pressure response during single-phase flow conditions; basic parameters are shown in Table 3.1, and the five-spot pattern is used as the calculation example. Figure 3.22 shows the pressure and derivative curves with well bore storage factors taking 0, 1, 5 and 10, respectively.

3.5.3. Influencing Characteristics of Oil/Water Viscosity Ratio on Pressure Response

The following presents the influence of different oil/water viscosity ratios on well testing pressure response during oil/water two-phase flow when well bore storage factor $C = 0.05\,\mathrm{m^3/MPa}$ and skin factor $S = 0.001$, and five-spot pattern is still used as the calculation example. Basic parameters are shown in Table 3.2. Figure 3.23 shows the pressure and derivative curves with oil/water viscosity ratios taking 5, 10 and 20, respectively.

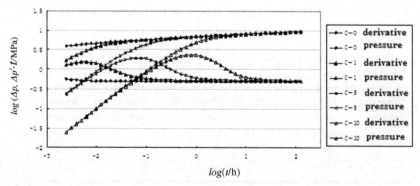

FIGURE 3.22   Pressure and derivative curves with different well bore storage factors.

TABLE 3.2 Basic Parameters

Reservoir area (m)	Oil formation thickness (m)	Porosity (%)	Permeability (μm^2)	Production rate (m^3/d)	Initial formation pressure (MPa)
630×630	10	30	800	40	19.2

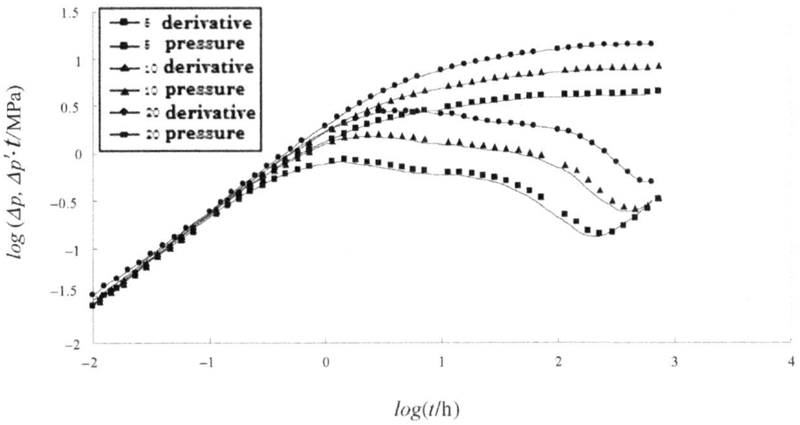

FIGURE 3.23 Pressure and derivative curves with different oil/water viscosity ratios.

It can be seen in Fig. 3.23 that there is an obvious concave in each pressure derivative curve. The fall of the pressure derivative is caused by the pseudo-constant pressure boundary formed by surrounding injection wells, and then the influence of the outer closed boundary on the testing well causes the upwarp of the pressure derivative curve. However, with the increase in oil/water viscosity ratio, the appearance of the concave in pressure derivative curve is earlier, since the pseudo-constant pressure boundary formed by surrounding water wells is closer to the production well.

3.5.4. Influencing Characteristic of Production History on Pressure Response

The following shows the influence of production history on well testing pressure response during oil/water two-phase flow when well bore storage factor $C = 0.5 \, m^3/MPa$ and skin factor $S = 0.1$, five-spot pattern is still taken as the calculation example. Basic parameters are shown in Table 3.3. Figure 3.24 presents the pressure and derivative curves with production time taking 30 d, 60 d, and 90 d, respectively.

TABLE 3.3 Basic Parameters

Reservoir area (m)	Oil formation thickness (m)	Porosity (%)	Permeability (μm^2)	Production rate (m^3 / d)	Initial formation pressure (MPa)
210×210	3	30	800	15	19.2

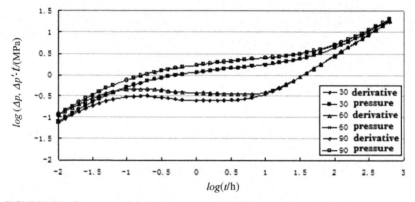

FIGURE 3.24   Pressure and derivative curves with different production history.

There are no concaves in the pressure derivative curves in Fig. 3.24. This is due to the small reservoir area and the relatively large production rate, and the oil/water surface has reached the production well when pressure build-up begins, so the pressure derivative curves could not reflect the influence of the pseudo-constant pressure boundary formed by the oil/water surface. Then the outer closed boundary affects the testing well and causes the upwarp of the derivative curves. With the increase in production time, the change in pressure and derivative curve decrease.

3.5.5. Influencing Characteristic of Well Pattern on Pressure Response
3.5.5.1. Inverted Nine-Spot Pattern
The following presents the characteristic curves of well testing pressure response in the inverted nine-spot pattern during oil/water two-phase flow with well bore storage factor C = 0.05 m^3 / MPa and skin factor S = 0.001. Drawdown test of injection well is taken as the calculation example and basic parameters are shown in Table 3.4. The streamline diagram, pressure and derivative curves of the system are shown in Figs 3.25 and 3.26.

TABLE 3.4 Basic Parameters

Reservoir area (m)	Oil formation thickness (m)	Porosity (%)	Permeability (μm^2)	Production rate (m^3 / d)	Initial formation pressure (MPa)
780×780	5	0.3	800	40	19.2

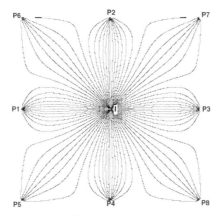

FIGURE 3.25   Streamline diagram of inverted nine-spot pattern system.

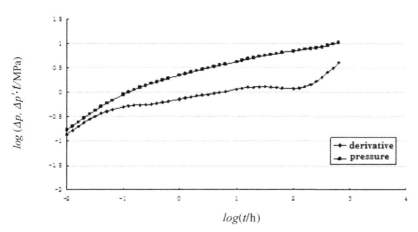

FIGURE 3.26   Pressure and derivative curves in water well drawdown test of inverted nine-spot pattern.

It can be seen from Fig. 3.26 that the derivative curve of the oil/water two-phase region (time log cycle 1~2) shows upwarp tendency and the peak point is the oil/water front. This curve is similar to the curve in the type-curve chart of injection well testing. Then the outer closed boundary influences the testing well and causes the upwarp of the pressure derivative curve.

3.5.5.2. Inverted Four-Spot Pattern

The following shows the characteristic curves of well testing pressure response in the inverted four-spot pattern during oil/water two-phase flow with well bore storage factor $C = 0.05\ m^3 / MPa$ and skin factor $S = 0.001$. Drawdown test of injection well is taken as the calculation example and basic parameters are shown in Table 3.5. The streamline diagram, pressure and derivative curves of the system are shown in Figs 3.27 and 3.28.

It can be seen from Fig. 3.28 that the derivative curve of the oil/water two-phase region (time log cycle 0~1) shows upwarp tendency and the peak point is the oil/water front. However, the degree of the upwarp tendency is small (due to the small well spacing), which reflects the oil/water front weakly. This curve is similar to the curve in the type-curve chart of injection well testing. Then the outer closed boundary influences the testing well and causes the upwarp of the pressure derivative curve.

TABLE 3.5 Basic Parameters

Reservoir area (m)	Oil formation thickness (m)	Porosity (%)	Permeability (μm^2)	Production rate (m^3 / d)	Initial formation pressure (MPa)
315×315	5	30	800	30	19.2

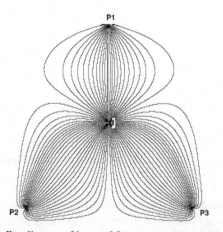

FIGURE 3.27 Streamline diagram of inverted four-spot pattern system.

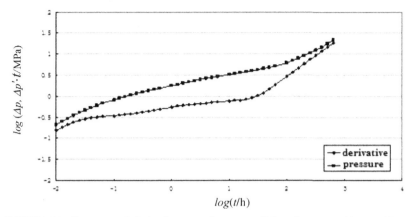

FIGURE 3.28   **Pressure and derivative curves in water well drawdown test of inverted four-spot pattern.**

3.5.5.3. Inverted Five-Spot Pattern

The following demonstrates the characteristic curves of well testing pressure response in the inverted five-spot pattern during oil/water two-phase flow with well bore storage factor $C = 0.05 \text{ m}^3/\text{MPa}$ and skin factor $S = 0.001$. Drawdown test of injection well is taken as the calculation example and basic parameters are shown in Table 3.6. The streamline diagram, pressure and derivative curves of the system are shown in Figs 3.29 and 3.30.

It can be seen from Fig. 3.30 that the pressure derivative curve of the oil/water two-phase region (time log cycle 1~2) shows upwarp tendency and the peak point is the oil/water front. This curve is similar to the curve in type-curve chart of injection well testing. Then the outer closed boundary influences the testing well and causes the upwarp of pressure derivative curve.

3.5.5.4. Nine-Spot Pattern

The following presents the characteristic curves of well testing pressure response in the nine-spot pattern during oil/water two-phase flow with well bore storage factor $C = 0.05 \text{ m}^3/\text{MPa}$ and skin factor $S = 0.001$. Drawdown test of production well is taken as the calculation example and basic parameters are

TABLE 3.6 Basic Parameters

Reservoir area (m)	Oil formation thickness (m)	Porosity (%)	Permeability (μm^2)	Production rate (m^3/d)	Initial formation pressure (MPa)
630×630	5	30	800	40	19.2

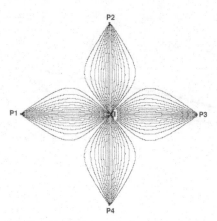

FIGURE 3.29   Streamline diagram of inverted five-spot pattern system.

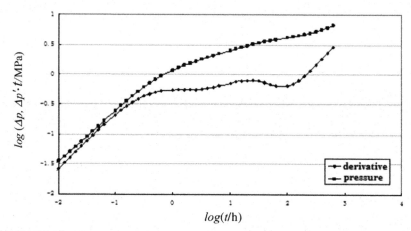

FIGURE 3.30   Pressure and derivative curves in water well drawdown test of inverted five-spot pattern.

shown in Table 3.7. The streamline diagram, pressure and derivative curves of the system are shown in Figs 3.31 and 3.32.

It can be seen from Fig. 3.32 that the derivative curve of the oil/water two-phase region (time log cycle 2∼3) shows a concave and the lowest point is the

TABLE 3.7 Basic Parameters

Reservoir area (m)	Oil formation thickness (m)	Porosity (%)	Permeability (μm^2)	Production rate (m^3/d)	Initial formation pressure (MPa)
780 × 780	5	30	800	40	19.2

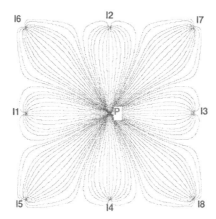

FIGURE 3.31 Streamline diagram of nine-spot pattern system.

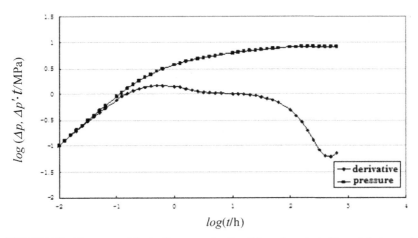

FIGURE 3.32 Pressure and derivative curves in oil well drawdown test of inverted nine-spot pattern.

oil/water front. This is because the injection rate of each water well is small, then the oil/water front is relatively far from the testing well, which reflects only a small segment on the derivative curve. The later closed boundary does not cause much upwarp tendency on the pressure derivative curve.

3.5.5.5. Row Well Pattern

The following shows the characteristic curves of well testing pressure response in row well pattern during oil/water two-phase flow with well bore storage factor $C = 0.05 \text{ m}^3 / \text{MPa}$ and skin factor $S = 0.001$. Pressure drawdown test of production well P2 is taken as the calculation example and basic parameters are shown in Table 3.7. The streamline diagram, pressure and derivative curves of the system are shown in Figs 3.33 and 3.34.

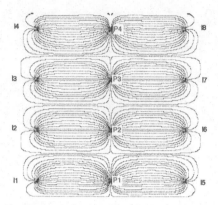

FIGURE 3.33   Streamline diagram of row well pattern.

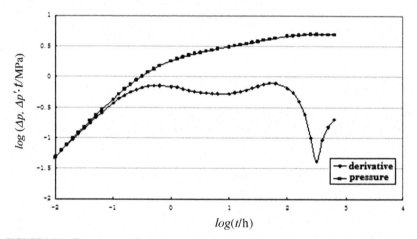

FIGURE 3.34   Pressure and derivative curves in oil well drawdown test of row well pattern.

It can be seen from Fig. 3.34 that the derivative curve of the oil/water two-phase region (time log cycle 2∼3) shows a severe concave and the lowest point of the concave is the oil/water front. Then the influence of surrounding production wells and closed boundary causes the upwarp tendency of the derivative curve.

3.5.5.6. Four-Spot Pattern

The following presents the characteristic curves of well testing pressure response in four-spot pattern during oil/water two-phase flow with well bore storage factor $C = 0.05 \, \text{m}^3 / \text{MPa}$ and skin factor $S = 0.001$. Drawdown test of production well is taken as the calculation example and basic parameters are shown in Table 3.8. The streamline diagram, pressure and derivative curves of the system are shown in Figs 3.35 and 3.36.

It can be seen from Fig. 3.36 that the derivative curve of the oil/water two-phase region (time log cycle 1∼2) shows a severe concave and the

TABLE 3.8 Basic Parameters					
Reservoir area (m)	Oil formation thickness (m)	Porosity (%)	Permeability (μm^2)	Production rate (m^3 / d)	Initial formation pressure (MPa)
315×315	5	30	800	30	19.2

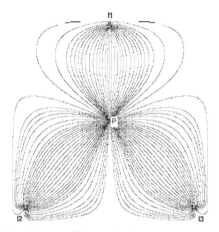

FIGURE 3.35   Streamline diagram of four-spot pattern.

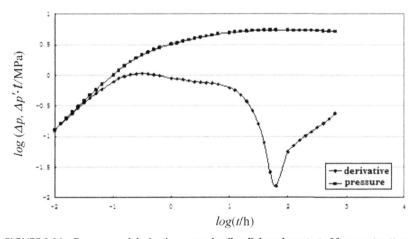

FIGURE 3.36   Pressure and derivative curves in oil well drawdown test of four-spot pattern.

lowest point is the oil/water front. The concave appears early because of the small well spacing. The later upwarp of derivative curve is caused by the surrounding closed boundary.

5.5.7. Five-Spot Pattern

The following shows the characteristic curves of well testing pressure response in the five-spot pattern during oil/water two-phase flow with well bore storage factor $C = 0.05 \, m^3 / MPa$ and skin factor $S = 0.001$. Pressure drawdown test of production well is taken as the calculation example and basic parameters are shown in Table 3.9. The streamline diagram, pressure and derivative curves of the system are shown in Figs 3.37 and 3.38.

It can be seen from Fig. 3.38 that the derivative curve of the oil/water two-phase region (time log cycle 2~3) is concave and the lowest point is the oil/water front. The later upwarp of derivative curve is caused by the surrounding closed boundary.

3.5.6. Influencing Characteristic of High Permeability Band Distribution on Pressure Response

Oil/water well two-spot and five-spot systems are taken as the calculation examples below; high-permeability bands are set along the main streamline direction, in shape S between the oil and water well, and perpendicular to the main streamline direction to simulate the real situation, then the

TABLE 3.9 Basic Parameters

Reservoir area (m)	Oil formation thickness (m)	Porosity (%)	Permeability (μm^2)	Production rate (m^3 / d)	Initial formation pressure (MPa)
630×630	5	30	800	40	19.2

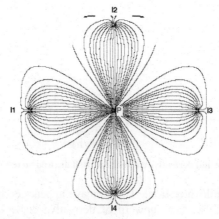

FIGURE 3.37   Streamline diagram of five-spot pattern.

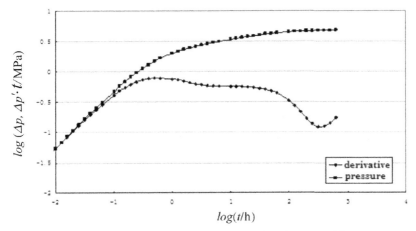

FIGURE 3.38   Pressure and derivative curves in oil well drawdown test of five-spot pattern.

influence of different highly permeable band distributions on pressure response are discussed.

3.5.6.1. Oil/Water Well Two-Spot Pattern
3.5.6.1.1. High Permeable Band Distribution Along The Main Streamline Direction

The following discusses the influence of high permeable band distribution along the main streamline direction on well testing pressure response during oil/water two-phase flow with skin factor $S = 0.001$ and well bore storage factor $C = 0.05 \text{ m}^3 / \text{MPa}$. Pressure drawdown test of the production well is taken as the calculation example and basic parameters are shown in Table 3.10. The permeability distribution diagram, streamline diagram, pressure and derivative curves are shown in Figs 3.39 to 3.41.

It can be seen from Fig. 3.41 that the pressure derivative curve in this case is obviously different from that of no high permeable band distribution. The pressure and derivative curves show a rising tendency; this is because the presence of high permeable band makes the transmission of pressure faster along the high permeable band than in other regions and then pressure drops

TABLE 3.10 Basic Parameters

Reservoir area (m)	Oil formation thickness (m)	Porosity (%)	Permeability (μm^2)	Production rate (m^3 / d)	Initial formation pressure (MPa)
1600×1600	5	30	800	40	19.2

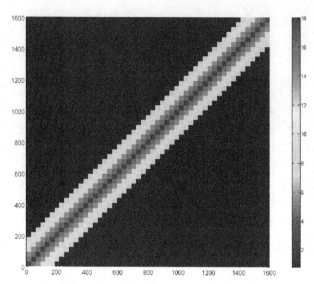

FIGURE 3.39 Permeability diagram of high permeable band distribution along main stream-line direction in oil/water well two-spot pattern.

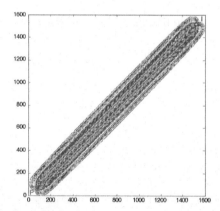

FIGURE 3.40 Streamline diagram of high permeable band distribution along main stream-line direction in oil/water well two-spot pattern.

quickly. In the region affected by high permeable band distribution (time log cycle 1~2), the pressure derivative curve shows a rising tendency, and the curve in the oil/water two-phase region (time log cycle 2.4~2.9) shows as concave.

3.5.6.1.2. High Permeable Band Distribution In S Shape

The following discusses the influence of high permeable band distribution in S shape on well testing pressure response during oil/water two-phase flow with skin factor $S=0.001$ and well bore storage factor $C=0.05\,m^3/MPa$. Pressure

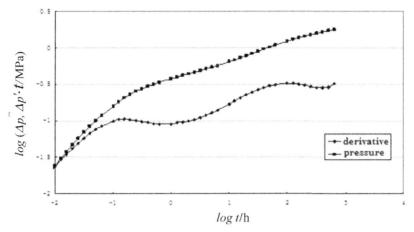

FIGURE 3.41 Pressure and derivative curves of high permeable band distribution along main streamline in oil/water well two-spot pattern.

drawdown test of the production well is taken as the calculation example and basic parameters are shown in Table 3.10. The permeability distribution diagram, streamline diagram, pressure and derivative curves are shown in Figs 3.42 to 3.44.

It can be seen from Fig. 3.44 that the pressure derivative curve in this case is also obviously different from that of no high permeable band distribution. The pressure and derivative curves also show a rising tendency; this is because the presence of high permeable band makes the transmission of pressure along the high permeable band faster than in other regions and then pressure drops

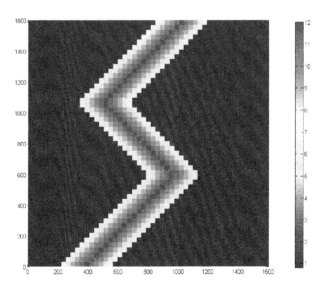

FIGURE 3.42 Permeability diagram of high permeable band distribution in S-shape in oil/water well two-spot pattern.

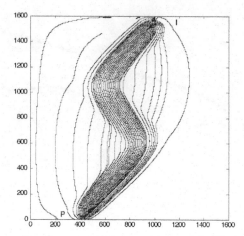

FIGURE 3.43   Streamline diagram of high permeable band distribution in S-shape in oil/ water well two-spot pattern.

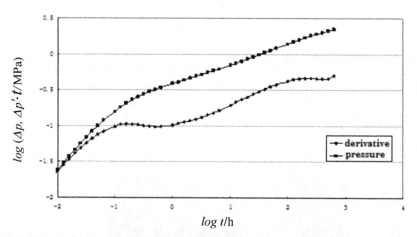

FIGURE 3.44   Pressure and derivative curves of high permeable band distribution in S shape in oil/water well two-spot pattern.

quickly. However, the influence on the oil/water two-phase region is not obvious. In the region affected by the high permeable band distribution (time log cycle 1.9~2.5), the pressure derivative curve is convex, and the curve in the oil/ water two-phase region (time log cycle 2.6~2.9) is concave.

3.5.6.1.3. High Permeable Band Distribution Perpendicular To Main Streamline Direction

The following discusses the influence of high permeable band distribution perpendicular to the main streamline direction on well testing pressure response during oil/water two-phase flow with skin factor S = 0.001 and well bore storage factor C = 0.05 m³/MPa. The pressure drawdown test of the production well is taken as the calculation example and basic parameters are shown in Table 3.10. The permeability distribution diagram, streamline diagram, pressure and derivative curves are shown in Figs 3.45 to 3.47.

In this case, the distribution of the high permeable band is perpendicular to the main streamline direction, and the pressure derivative curve is similar to that of no high permeable band distribution. The main reason is that the presence of high permeable band does not have a great influence on pressure transmission. It can also be seen from the streamline diagram that the streamline shape does not change much. The influence of the oil/water front is not reflected due to the big well spacing, and the pressure derivative curve does not show as concave.

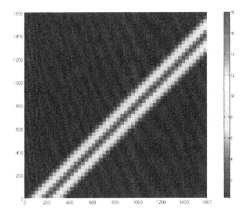

FIGURE 3.45   Permeability diagram of high permeable band distribution perpendicular to main streamline direction in oil/water well two-spot pattern.

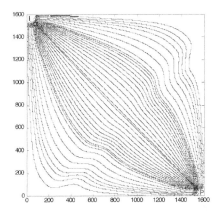

FIGURE 3.46   Streamline diagram of high band distribution perpendicular to main streamline direction in oil/water well two-spot pattern.

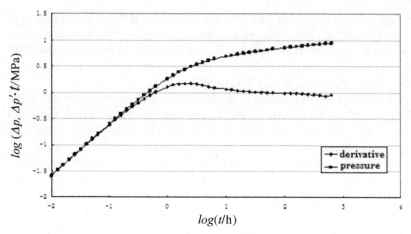

FIGURE 3.47   Pressure and derivative curves of high permeable band distribution perpendicular to main streamline direction in oil/water well two-spot pattern.

3.5.6.2.   Oil/water Well Five-Spot Pattern
3.5.6.2.1.   High Permeable Band Distribution Along Main Streamline Direction

The following discusses the influence of high permeable band distribution along the main streamline direction on well testing pressure response during the oil/water two-phase flow with skin factor $S = 0.001$ and well bore storage factor $C = 0.05 \, m^3 / MPa$. The pressure drawdown test of the production well is taken as the calculation example and basic parameters are shown in Table 3.11. The permeability distribution diagram, streamline diagram, pressure and derivative curves are shown in Figs 3.48 to 3.50.

It can be seen from Fig. 3.50 that the pressure derivative curve in this case is obviously different from that of no high permeable band distribution. The pressure and derivative curves show rising tendencies; this is because the presence of high permeable band makes the transmission of pressure faster along the high permeable band than in other regions, and then pressure drops quickly. In the region affected by the high permeable band distribution (time log cycle

TABLE 3.11 Basic Parameters

Reservoir area (m)	Oil formation thickness (m)	Porosity (%)	Permeability (μm^2)	Production rate (m^3 / d)	Initial formation pressure (MPa)
1600 × 1600	5	30	800	40	19.2

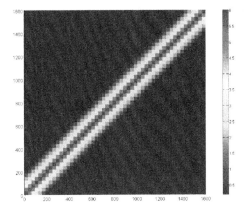

FIGURE 3.48   Permeability diagram of high permeable band distribution along main streamline direction in five-spot pattern.

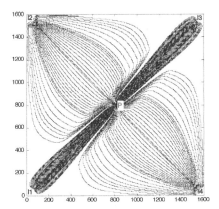

FIGURE 3.49   Streamline diagram of high permeable band distribution along main streamline direction in five-spot pattern.

1.6~2.3), the pressure derivative curve is convex, and the influence of the other two infection wells makes the convex zone wider; the curve in the oil/water two-phase region (time log cycle 2.4~2.9) is concave.

3.5.6.2.2. High Permeable Band Distribution in S Shape

The following discusses the influence of high permeable band distribution in S shape on well testing pressure response during the oil/water two-phase flow with skin factor $S = 0.001$ and well bore storage factor $C = 0.05 \, \text{m}^3 / \text{MPa}$. The pressure drawdown test of the production well is taken as the calculation example and basic parameters are shown in Table 3.11. The permeability distribution diagram, streamline diagram, pressure and derivative curves are shown in Figs 3.51 to 3.53.

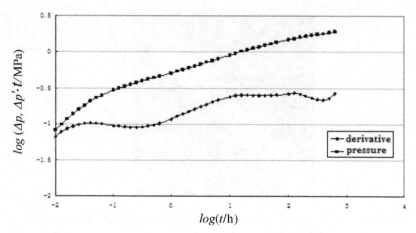

FIGURE 3.50   Pressure and derivative curves of high permeable banded distribution along main streamline direction in five-spot pattern.

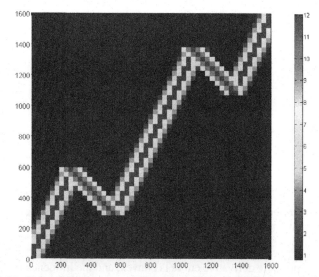

FIGURE 3.51   Permeability diagram of high permeable band distribution in S-shape in five-spot pattern.

It can be seen from Fig. 3.53 that the pressure derivative curve in this case is also obviously different from that of no high permeable band distribution. The pressure and derivative curves also show rising tendency; this is because the presence of high permeable band makes the transmission of pressure along the high permeable band faster than in the other regions, and then pressure drops quickly. In the region affected by high permeable band distribution (time log cycle 1.9~2.5), the pressure derivative curve is convex, and the curve in the oil/water two-phase region (time log cycle 2.2~2.9) is concave.

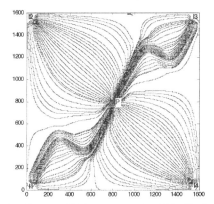

FIGURE 3.52   Streamline diagram of high permeable band distribution in S shape in five-spot pattern.

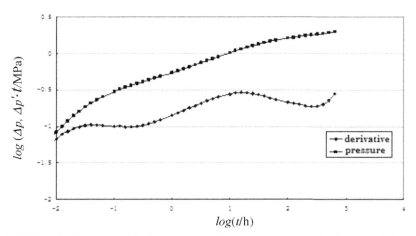

FIGURE 3.53   Pressure and derivative curves of high permeable band distribution in S shape in five-spot pattern.

3.5.6.2.3. High Permeable Band Distribution Perpendicular to Main Streamline Direction

The following discusses the influence of high permeable band distribution perpendicular to the main streamline direction on well testing pressure response during the oil/water two-phase flow with skin factor $S = 0.001$ and well bore storage factor $C = 0.05 \, \text{m}^3 / \text{MPa}$. Pressure drawdown test of production well is taken as the calculation example and basic parameters are shown in Table 3.11. The permeability distribution diagram, streamline diagram, pressure and derivative curves are shown in Figs 3.54 to 3.56.

In this case, the distribution of high permeable band is perpendicular to the main streamline direction, and the pressure derivative curve is similar to that of no high permeable band distribution. The main reason is that the presence of

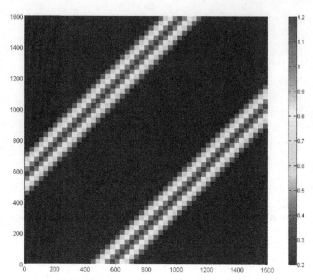

FIGURE 3.54   Permeability diagram of high permeable band distribution perpendicular to main streamline direction in five-spot pattern.

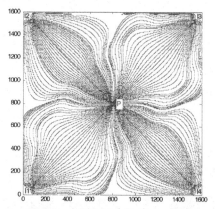

FIGURE 3.55   Streamline diagram of high permeable band distribution perpendicular to main streamline direction in five-spot pattern.

high permeable band does not have a great influence on pressure transmission. It can also be seen from the streamline diagram that the streamline shape does not change much, and the influence of the oil/water front makes the pressure derivative curve concave.

3.6. CHAPTER SUMMARY

In this chapter, the streamline numerical well testing interpretation model and pressure response characteristics for a single-layer sandstone waterflooding reservoir are studied. The main work includes:

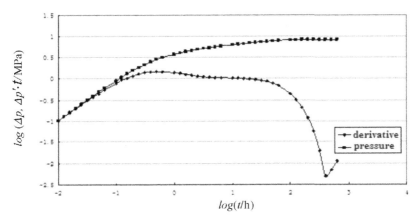

FIGURE 3.56   Pressure and derivative curves of high permeable band distribution perpendicular to main streamline direction in five-spot pattern.

1. The numerical well testing interpretation model for a single-layer sandstone waterflooding reservoir includes filtration mathematical model of the production period and the streamline mathematical model of the testing period. A filtration mathematical model of the production period is used to simulate the production history, and to obtain the pressure, saturation and streamline distribution at the time of well test. The streamline mathematical model of testing period is used to simulate the pressure variation process of the testing well in the testing period, and to obtain theoretical pressure response of the testing well. Pressure, saturation, and streamline distribution calculated in the production period are taken as the initial conditions for the establishment of the streamline mathematical model in the testing period.

2. The filtration mathematical model of the production period applies the simplified black oil model, which considers many factors including heterogeneity, oil/water two-phase flow, complex boundary and multi-well production. So the model accords with the actual reservoir well. The combination of the IMPES and streamline methods is used to solve the filtration mathematical model of the production period; namely, the IMPES method is used to solve the pressure distribution while the streamline method is used to solve the saturation distribution. The concrete procedures are as follows: first, the IMPES method is used to solve pressure control equation and to obtain pressure distribution of the research region; then the streamlines are traced based on pressure distribution and the streamline distribution is obtained; the grid parameters are converted to streamline node parameters and saturation equations are established along the streamlines; last, the saturation equations of all streamlines are solved and the saturation distribution of the research region is obtained.

There is no need to calculate the pressure at every time-step when solving saturation with the streamline method (unless events that could change the streamline distribution occur). The solution of saturation and pressure are separated, which increases the solution speed greatly. Also the saturation equation along the streamline is one-dimensional, so the computation has better stability with slight numerical dispersion and easily controlled error. The mathematical model of the testing period is the streamline model, which is made up of a set of mathematical equations that are established by each streamline. The mathematical equation along each streamline could consider the variation of reservoir geology and fluid parameters. A simultaneous method is used to solve the streamline mathematical models of the testing period, which can consider the influence of each streamline equation on the pressure response of the testing well.

3. The validity of the streamline numerical well testing interpretation model for a single-layer sandstone waterflooding reservoir is verified from the resolution of three typical models.

   a. For the typical model of homogeneous infinite reservoir (single-phase flow), pressure response after the testing well shut-in is simulated, the calculation results are completely consistent with that of analytical solution, which verifies the validity of the streamline numerical well testing interpretation model and its solving method in single-phase flow conditions.

   b. For the typical model of a homogeneous waterflooding reservoir (two-phase flow), pressure response after the testing well shut-in is simulated, the calculation results correspond with the pressure response characteristics of oil/water two-phase flow, which verifies the validity of the streamline numerical well testing interpretation model in two-phase flow conditions.

   c. For the typical model of a heterogeneous reservoir (single-phase flow), pressure response after the testing well shut-in is simulated, calculation results correspond with the pressure response characteristics of a heterogeneous reservoir, which verifies the validity of the streamline numerical well testing interpretation model in heterogeneous conditions.

Streamline Numerical Well Testing Interpretation Model for a Multi-Layer Sandstone Waterflooding Reservoir

Multi-layer reservoir well testing is always the hot problem in well testing interpretation theory and methods. Since the birth of numerical well testing, much research on multi-layer reservoir numerical well testing has been done at home and abroad, which mainly obtains the pressure response of the numerical well testing interpretation model by coupling the production model with the flow control equation. However, most of these models are based on two-dimensional or three-dimensional complex grid systems, so the solving speed and stability make it difficult to satisfy the requirements of well testing interpretation. In this chapter, based on the streamline numerical well testing interpretation model for a single-layer reservoir described in the previous chapter, the streamline numerical well testing interpretation model for a multi-layer reservoir is established and solved, pressure response characteristics of the streamline numerical well testing interpretation model for a multi-layer reservoir with cross-flow and no cross-flow are analyzed.

4.1. THE BUILDING OF A STREAMLINE NUMERICAL WELL TESTING INTERPRETATION MODEL FOR A MULTI-LAYER SANDSTONE WATERFLOODING RESERVOIR

The streamline numerical well testing interpretation model for a multi-layer waterflooding reservoir includes a filtration mathematical model of the production period and a streamline mathematical model of the testing period.

4.1.1. Filtration Mathematical Model of the Production Period

The filtration mathematical model of the production period for a multi-layer waterflooding reservoir is similar to that of a single-layer waterflooding

Streamline Numerical Well Test Interpretation. DOI: 10.1016/B978-0-12-386027-9.00004-6
Copyright © 2011 by Elsevier Ltd

reservoir, and both of them use a simplified black oil model; the difference is that the multi-layer reservoir model needs to consider the assignment of the production term in each layer.

The filtration mathematical model of the production period for a multi-layer waterflooding reservoir is based on the following assumptions:

1. the fluids in reservoir are water and oil;
2. the reservoir is a multi-layer reservoir, and oil/water wells are commingled production or injection wells;
3. rocks and fluids in the reservoir are incompressible;
4. fluid flow in reservoir obeys Darcy's law;
5. the reservoir is heterogeneous and the permeability is isotropic in the plane and anisotropic in the vertical; and
6. capillary and gravity force are ignored.

The filtration mathematical model of the production period for a multi-layer waterflooding reservoir is shown below.

Pressure control equation of oil phase:

$$\nabla \cdot \left(\frac{\rho_o K K_{ro}}{\mu_o} \nabla p \right) + q_o = \frac{\partial (\phi \rho_o S_o)}{\partial t} \tag{4.1.1}$$

Pressure control equation of water phase:

$$\nabla \cdot \left(\frac{\rho_w K K_{rw}}{\mu_w} \nabla p \right) + q_w = \frac{\partial (\phi \rho_w S_w)}{\partial t} \tag{4.1.2}$$

Saturation normalization equation:

$$S_o + S_w = 1 \tag{4.1.3}$$

Initial condition:

$$\begin{cases} p|_{t=0} = p_i \\ S_w|_{t=0} = S_{wi} \end{cases} \tag{4.1.4}$$

The outer boundary condition is the same as that of the well testing interpretation model for a single-layer reservoir.

Inner boundary condition, which is the issue of well processing in a constant liquid production rate condition, is considered here.

In the filtration mathematical model of the production period for a multi-layer reservoir, a well will cross through several grids in the vertical. The surface conditions given are all for the total well, which means the production rate is the sum of the output of all the grids crossed by the well. However, the rate in differential equations is relevant to each concrete grid, so it refers not only to the rate distribution problem among different phases but also to the rate distribution problem among different grids. The processing method is described as follows.

In an xyz three-dimensional space system, if z is the direction of the hole axis, do not consider the change of fluid pressure in well bore; the PID index of well gridblocks is defined as

$$PID = \frac{2\pi\Delta z \sqrt{K_x K_y}}{\ln\frac{r_e}{r_w} - 0.75} \qquad (4.1.5)$$

Since the permeability is isotropic in the plane, which means that $K_x = K_y = K_h$, then Equation 4.1.5 can be written as

$$PID = \frac{2\pi\Delta z K_h}{\ln\frac{r_e}{r_w} - 0.75} \qquad (4.1.6)$$

For rectangular grids in vertical wells,

$$r_e = 0.14\sqrt{\Delta x^2 + \Delta y^2} \qquad (4.1.7)$$

In the above formulae, K_x and K_y are the permeability of well grids in x and y directions, respectively (μm^2); K_h is the permeability of well grids in the horizontal direction (μm^2); Δx, Δy and Δz are the grid steps of well grids in x, y and z directions, respectively (cm); r_e is the equivalent supply radius (cm).

Suppose that M simulative grid blocks (completion section) are perforated by the well, in the condition of constant fluid production, for the production well:

Total oil production rate of all completion sections:

$$Q_o = \frac{\sum\limits_{m=1}^{M} (PID\lambda_o)_m}{\sum\limits_{m=1}^{M} (PID\lambda_o)_m + \sum\limits_{m=1}^{M} (PID\lambda_w)_m} Q_l \qquad (4.1.8)$$

Oil production rate of each completion section:

$$Q_{om} = \frac{(PID\lambda_o)_m}{\sum\limits_{m=1}^{M} (PID\lambda_o)_m} Q_o \qquad (4.1.9)$$

Water production rate of each completion section:

$$Q_{wm} = \frac{\lambda_{wm}}{\lambda_{om}} Q_o \qquad (4.1.10)$$

For an injection well with constant injection rate:

$$Q_{lm} = \frac{[PID(\lambda_o + \lambda_w)]_m}{\sum\limits_{m=1}^{M} [PID(\lambda_o + \lambda_w)]_m} Q_l \qquad (4.1.11)$$

4.1.2. The Streamline Mathematical Model of the Testing Period

Make superscript m represent layer order, subscript j represent streamline order. Suppose the streamline number emitting from the testing well in layer m is N^m, then filtration control equations of all streamlines in M layers are:

$$\frac{1}{l_j^m}\frac{\partial}{\partial l_j^m}\left(\alpha l_j^m \frac{\lambda_j^m}{\phi_j^m}\frac{\partial p_j^m}{\partial l_j^m}\right) = \alpha c_t \frac{\partial p_j^m}{\partial t}\ (m=1,2,\cdots,M)\,(j=1,2,\cdots,N^m)$$

(4.1.12)

where, l_j^m is the curvilinear coordinate of jth streamline in layer m and the origin coordinate is the location of the testing well, cm; p_j^m is the pressure of jth streamline in layer m, 10^{-1}MPa; ϕ_j^m and λ_{tj}^m are the porosity and total mobility of the reservoir crossed by jth streamline in layer m, respectively [dimensionless and $\mu m^2/(mPa\bullet s)$].

For the formation with isotropic permeability in the plane and anisotropic permeability in the vertical, the absolute permeability along any direction of the streamline is calculated as follows:

$$K = \sqrt{(K_h\cos\theta)^2 + (K_v\sin\theta)^2}$$

(4.1.13)

where, K_h and K_v are the permeability in the plane and the vertical, respectively (μm^2); θ is the included angle between streamline and horizontal direction.

Inner boundary condition: well bore storage effect and skin effect are considered. For well bore storage inner boundary, the testing well is separated into two situations: production well and injection well.

Well bore storage inner boundary condition of the production well:

$$\sum_{m=1}^{M}\sum_{j=1}^{N^m}\frac{2\pi}{N^m}r_w h_1^m\left(\lambda_{tj}^m\frac{\partial p_j^m}{\partial l_j^m}\right)_{l_j^m=r_w}$$

$$= -q + C\frac{dp_{wf}}{dt}\ (m=1,2,\cdots,M)\,(j=1,2,\cdots,N^m)$$

(4.1.14)

Well bore storage inner boundary condition of the injection well:

$$\sum_{m=1}^{M}\sum_{j=1}^{N^m}\frac{2\pi}{N^m}r_w h_1^m\left(\lambda_{tj}^m\frac{\partial p_j^m}{\partial l_j^m}\right)_{l_j^m=r_w}$$

$$= q + C\frac{dp_{wf}}{dt}\ (m=1,2,\cdots,M)\,(j=1,2,\cdots,N^m)$$

(4.1.15)

Skin inner boundary condition (simultaneous condition of well bottom flow pressure):

$$p_{wf} = p_{wj}^m - S^m\left(l_j^m\frac{\partial p_j^m}{\partial l_j^m}\right)_{l_j^m=r_w}\quad (m=1,2,\cdots,M)\,(j=1,2,\cdots,N^m)\quad (4.1.16)$$

Underground flow rate of each layer could be calculated by the following formula:

$$q_m = \sum_{j=1}^{N^m} \frac{2\pi}{N^m} r_w h_1^m \left(\lambda_{tj}^m \frac{\partial p_j^m}{\partial l_j^m} \right)_{l_j^m = r_w} \quad (m = 1, 2, \cdots, M) \, (j = 1, 2, \cdots, N^m)$$

(4.1.17)

where, p_{wj}^m is the first node pressure of jth streamline in layer m, 10^{-1}MPa; h_1^m is the effective thickness of layer m, cm; S^m is the skin factor of layer m (dimensionless).

Outer boundary condition: this is the same as that of the streamline mathematical model of the testing period for a single-layer sandstone waterflooding reservoir.

Initial condition: this is obtained by solving the filtration mathematical model of the production period.

Saturation equation:

$$\frac{\partial S_w}{\partial t} + \frac{\partial f_w}{\partial \tau} = 0$$

(4.1.18)

4.2. SOLUTION OF THE STREAMLINE NUMERICAL WELL TESTING INTERPRETATION MODEL FOR A MULTI-LAYER SANDSTONE WATERFLOODING RESERVOIR

4.2.1. Solving Method of the Filtration Mathematical Model of the Production Period

The solving method of the filtration mathematical model of the production period for a multi-layer waterflooding reservoir is the same as that of a single-layer sandstone waterflooding reservoir. Four procedures are mainly included:

Firstly, use the IMPES method to solve the pressure control equation and obtain the pressure and saturation distribution of the multi-layer reservoir grid system.

Secondly, use the Pollock method to trace the streamline based on the pressure and saturation distribution.

Thirdly, convert the grid parameters to node parameters of each streamline, solve the one-dimensional saturation equation along the streamline, and obtain the saturation distribution along each streamline of the testing well at the time of well testing.

Finally, convert the saturation parameters of all streamline nodes into the corresponding grids using TOF weighting method, and update the streamlines to the time of well testing.

If events which can greatly change the streamline distribution or the node parameters occur in the production history, streamlines should be updated according to time.

2.2. Solving Method of the Streamline Mathematical Model of the Testing Period

The solving method of the streamline mathematical model in the testing period for a multi-layer sandstone waterflooding reservoir is similar to that of a single-layer sandstone waterflooding reservoir; namely, diverge the inner boundary condition Equations 4.1.14 to 4.1.16 and obtain differential Equation 4.2.1; diverge pressure control Equation 4.1.12 and obtain differential Equation 4.2.2; diverge outer boundary condition equation and obtain Equation 4.2.3; then obtain the differential equations of each streamline:

$$a_{ji}p_{ji}^{n+1} + c_{ji}p_{ji+1}^{n+1} = g_{ji} \ (i=1) \tag{4.2.1}$$

$$b_{ji}p_{ji-1}^{n+1} + a_{ji}p_{ji}^{n+1} + c_{ji}p_{ji+1}^{n+1} = g_{ji} \ (i=2,\cdots,N_j) \tag{4.2.2}$$

$$b_{ji}p_{ji-1}^{n+1} + a_{ji}p_{ji}^{n+1} = g_{ji} \ (i=N_j) \tag{4.2.3}$$

The difference is that the solution of a multi-layer reservoir model needs to solve all the streamline mathematical equations of all layers that are relevant to the testing well together. For the testing well with M layers perforated, if the streamline number of each layer is N_m, and N_J nodes are included in each streamline, then the total number of the simultaneous equations for the streamline mathematical model is $\sum_{m=1}^{M}\sum_{j=1}^{N_m}N_j$. The solving method of saturation equations is the same as that of a single-layer reservoir model.

The coefficient matrix of the differential equations in the streamline mathematical model of the testing period for a multi-layer waterflooding reservoir is

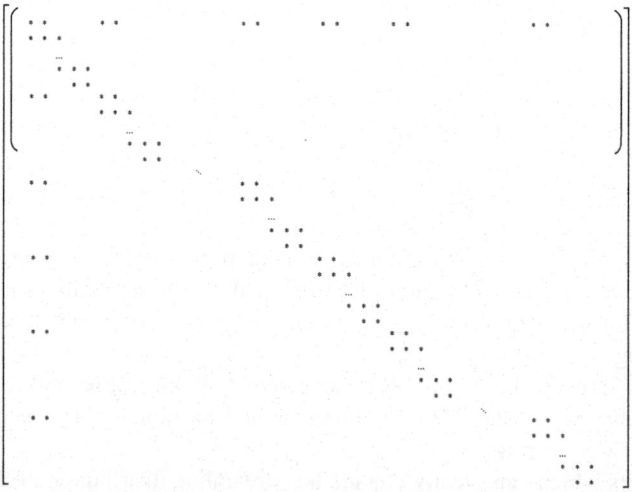

FIGURE 4.1   Coefficient matrix structure of differential equations of streamline mathematical model for a two-layer reservoir.

a sparse matrix with the size of $\sum_{m=1}^{M}\sum_{j=1}^{N_m}N_j \times \sum_{m=1}^{M}\sum_{j=1}^{N_m}N_j$. Figure 4.1 shows the coefficient matrix structure of differential equations of streamline mathematical model for a two-layer reservoir, the small partitioned matrix is the coefficient matrix of simultaneous differential equations for the first layer in the streamline mathematical model.

4.3. PRESSURE RESPONSE CHARACTERISTICS OF THE STREAMLINE NUMERICAL WELL TESTING INTERPRETATION MODEL FOR A MULTI-LAYER SANDSTONE WATERFLOODING RESERVOIR

4.3.1. Influencing Characteristic of Well Bore Storage Factor on Pressure Response in the Well Testing Interpretation Model

In order to study the influencing characteristic of well bore storage factor on pressure response of the streamline numerical well testing interpretation model for a multi-layer waterflooding reservoir with cross-flow and no cross-flow, a typical model of an inverted five-spot pattern in a three-layer homogeneous reservoir is established. The main parameters are described below.

Permeability is different for each reservoir plane, and the value of 1st, 2nd and 3rd layers (from the top down) are 0.01 μm^2, 0.1 μm^2 and 1 μm^2, respectively. Porosity and effective thickness are the same for each layer: porosity is 0.2 and effective thickness is 1 m. Physical property parameters and phase permeability are the same for oil and water phases, so the model is degenerated into a single-phase flow model. The fluid production rate of oil wells P1~P3 is 80 m^3/d, fluid production rate of oil well P4 is 60 m^3/d, fluid injection rate of water well I is 200 m^3/d. The 1st, 2nd and 3rd layers of wells P1~P4 and I are all perforated. The plane distribution of the well location is shown in Fig. 4.2.

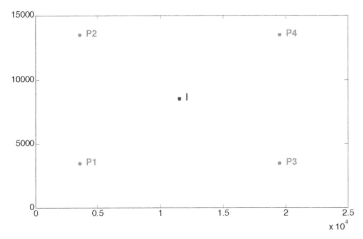

FIGURE 4.2 **Plane distribution of well location.**

Streamline distribution without and with cross-flow are shown in Figs 4.3 and 4.4, respectively. In the conditions without and with interlayer cross-flow, injection well I is shut in and different well bore storage coefficients are taken to simulate the pressure drawdown in testing well I, then obtain the pressure response curves of the two different cases as shown in Figs 4.5 and 4.6.

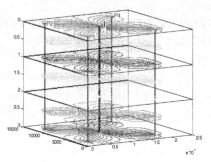

FIGURE 4.3   Streamline distribution without cross-flow.

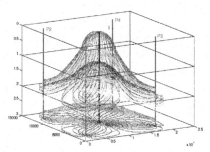

FIGURE 4.4   Streamline distribution with cross-flow.

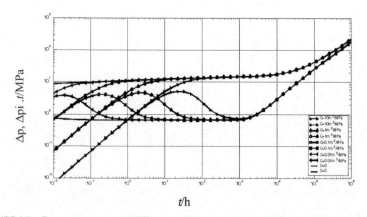

t/h

FIGURE 4.5   Pressure response of different well bore storage coefficients without cross-flow.

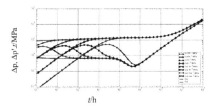

FIGURE 4.6   Pressure response of different well bore storage coefficients with cross-flow.

It can be seen from Fig. 4.6 that a concave phenomenon appears in the middle and late period of pressure derivative, which is caused by the cross-flow from the low permeability layer to the high permeability layer. As seen in Fig. 4.4, when streamlines reach the neighboring wells in the middle and late periods, streamlines enter into the high permeability layer from the low permeability layer. Without considering cross-flow, streamlines stay in one layer (as seen in Fig. 4.3) and there is no change of mobility, so the concave phenomenon of pressure derivative will not happen (as seen in Fig. 4.4).

It can also be seen from Figs 4.4 and 4.5 that, when considering well bore storage effect in the well testing interpretation model, with the increase of well bore storage factor, the whole pressure response curve moves to the right, and the appearance of each flow period is postponed. When considering inter-layer cross-flow, the influence of the well bore storage effect on the start time of inter-layer cross-flow is small, but too large a well bore storage factor may cover the appearance of cross-flow.

4.3.2. The Influencing Characteristic of Skin Factor on Pressure Response in the Well Testing Interpretation Model

Based on the above typical model, the influencing characteristic of skin factor on the streamline numerical well testing interpretation model for a multi-layer waterflooding reservoir is studied.

At first, the same skin factor is taken for each layer. For different skin factors, pressure response curves without and with inter-layer cross-flow are obtained and shown in Figs 4.7 and 4.8, respectively. Then change the skin factor of the middle layer and fix the skin factors of the other layers, the pressure response curves without and with inter-layer cross-flow are obtained and shown in Figs 4.9 and 4.10.

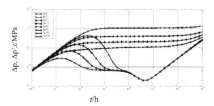

FIGURE 4.7   Pressure response with cross-flow when skin factors of each layer are same.

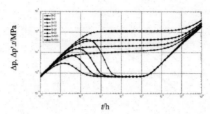

FIGURE 4.8   Pressure response without cross-flow when skin factors of each layer are same.

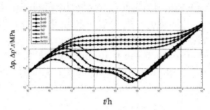

FIGURE 4.9   Pressure response with cross-flow when skin factors of each layer are different.

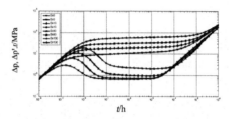

FIGURE 4.10   Pressure response without cross-flow when skin factors of each layer are different.

It can be seen from Figs 4.7 to 4.10 that when skin factors of each layer in a multi-layer reservoir are partly or totally changed, on the one hand, with the increase of skin factor, well bore damage becomes serious and higher additional pressure drop occurs, which increases the peak value of pressure derivative in the early period; on the other hand, with the increase of skin factor, the damage zone broadens and the time of fluid flowing through the damage zone increases, which delays the time of pressure response entering the radial flow period.

It can be seen from Figs 4.7 and 4.8 that when skin factors of each layer are equivalent and change the skin factor in each layer at the same time, the pressure derivative curves in the middle radial flow period and the later flow period are almost coincidental, which indicates that the influence of the same skin factor on each layer is the same. As seen in Figs 4.9 and 4.10, when skin factor of some layer in a multi-layer reservoir is partly changed and then the skin factors of each layer are different, the pressure derivative curves in radial flow and later flow period do not coincide, which is because different skin factors cause different additional pressure drops in each layer.

4.3.3. Influencing Characteristics of Vertical and Horizontal Permeability Ratio on Pressure Response in the Well Testing Interpretation Model

Based on the above typical model, the influencing characteristic of vertical and horizontal permeability ratio on streamline numerical well testing interpretation model for multi-layer waterflooding reservoir is studied.

Suppose the permeability ratios in each layer are equivalent, pressure response curves are obtained through calculation with the permeability ratios taking 0, 0.001, 0.01, 0.1, 1 and 10, respectively (shown in Fig. 4.11).

As can be seen in Fig. 4.11, when permeability ratio is 0 (vertical permeability is 0), no inter-layer cross-flow occurs, pressure derivative curve is a horizontal straight-line segment in the middle radial flow region. When permeability ratio is not 0, there is a "concave" in the pressure derivative curve, and with the increase of permeability ratio: (1) the earlier the time of pressure derivative concave appears, which means cross-flow occurs earlier; (2) the deeper the pressure derivative concave becomes, which means the inter-layer cross-flow is more serious.

4.3.4. Characteristics of the Streamline Numerical Well Testing Interpretation Model for a Multi-Layer Waterflooding Reservoir with Complex Cross-Flow

There is a common characteristic of inter-layer cross-flow in the above research, i.e., the streamlines are always from low permeability layers to high permeability layers, which leads to the concave on the pressure derivative curve. The one-way cross-flow phenomenon from a low permeability layer to a high permeability

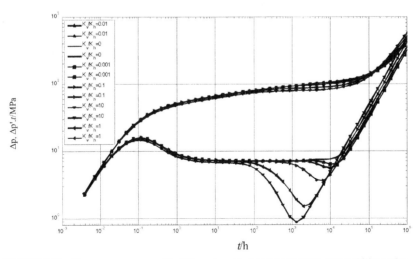

FIGURE 4.11 Pressure response with different vertical and horizontal permeability ratios.

layer is obtained on the condition that the permeability is homogeneous in each layer and the layers of neighboring wells are all perforated. In this case, reservoir pressure depletion in each layer at the production well is directly proportional to the layer permeability, pressure of high permeability layer at the production well is low, and so the fluid will flow from the low permeability layer to the high permeability layer. However, cross-flow in real conditions is very complex, the cross-flow exists not only from the low permeability layer to the high permeability layer, but also from the high permeability layer to the low permeability layer.

In order to study the characteristics of the streamline numerical well testing interpretation model for a multi-layer waterflooding reservoir with complex cross-flow, a typical model of an inverted five-spot pattern in a three-layer reservoir is built and the main parameters are shown below.

Permeability is homogeneous in each layer and heterogeneous among different layers, and the horizontal permeability of 1st, 2nd, 3rd layers (from the top down) are $2.4 \times 10^{-3} \mu m^2$, $0.4 \times 10^{-3} \mu m^2$ and $0.2 \times 10^{-3} \mu m^2$, respectively. Porosity and effective thickness are the same for each layer, porosity is 0.2 and effective thickness is 50 m. Oil and water phases have the same physical parameters and phase permeabilities, so the model is degenerated into a single-phase flow model. Completion segments and production performance of each well are shown in Table 4.1, and plane distribution of well location is shown in Fig. 4.12.

In order to study the pressure response characteristics of the streamline numerical well testing interpretation model for a multi-layer waterflooding reservoir with complex cross-flow, pressure response characteristic curves without and with cross-flow are calculated, respectively.

Figure 4.13 shows that, in the case without cross-flow, fluid flows in the plane. Because layer permeability and perforated position are different, the streamline distribution of each layer is different, and the high fluid injection

TABLE 4.1 Completion Segments and Production Performance of Each Well

		Parameter	
Well name	Perforated layers	Well type	Production/injection rate (m^3/d)
P1	1	Production well	20
P2	2	Production well	20
P3	3	Production well	20
P4	1, 3	Production well	20
I	1, 2, 3	Injection well	− 100

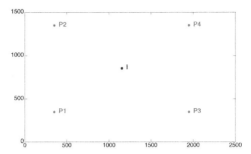

FIGURE 4.12   Plane distribution of well location.

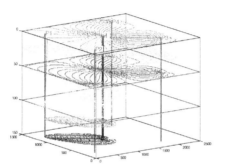

FIGURE 4.13   Streamline distribution without cross-flow.

and production rate of the high permeability layer leads to high pressure drop and large affected area of streamlines. The streamline diagrams with cross-flow are shown in Figs 4.14 and 4.15. Figure 4.15 shows the streamline distribution from one horizontal plane; it can be seen that the cross-flow is very complex in this case, the cross-flow exists not only from the low permeability layer to the high permeability layer, but also from the high permeability layer to the low permeability layer. However, due to the difference of inter-layer permeability and production zone of the production well, the cross-flow from the high permeability layer to the low permeability layer is predominant.

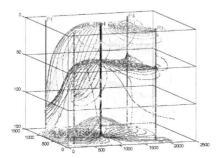

FIGURE 4.14   Streamline distribution with cross-flow.

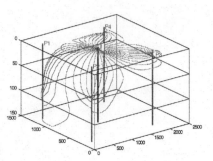

FIGURE 4.15   Streamline distribution from one horizontal plane with cross-flow.

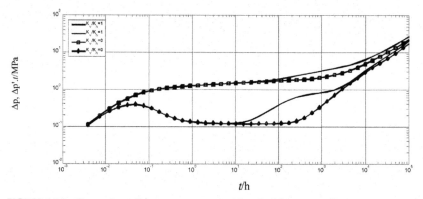

FIGURE 4.16   Comparison of pressure response with and without cross-flow.

It can be seen from Figs 4.14 to 4.16 that the asymmetry of the fluid production rate created by production well perforation causes the complex interlayer cross-flow, and the cross-flow exists not only from the low permeability layer to the high permeability layer, but also from the high permeability layer to the low permeability layer. However, the injection rate of the high permeability layer is higher and the pressure is higher than the low permeability layer, so the cross-flow from the high permeability layer to the low permeability is predominant. When cross-flow happens from the high permeability layer to the low permeability layer, the pressure derivative curve shows upwarp.

4.3.5. Characteristics of the Streamline Numerical Well Testing Interpretation Model for a Multi-Layer Sandstone Waterflooding Reservoir with Oil/Water Two-Phase Flow

In order to study the pressure response characteristics of the streamline numerical well testing interpretation model for a multi-layer sandstone waterflooding reservoir with oil/water two-phase flow, a model of a typical three-layer reservoir with constant pressure outer boundary is established. Perforated zones

TABLE 4.2 Main Parameters of Each Well

		Parameter	
Well name	Perforated layers	Well type	Production/injection rate (m³ / d)
P1	1, 2	Production well	60
P2	2, 3	Production well	60
P3	1, 3	Production well	60
P4	1, 2	Production well	60
P5	3	Production well	60
I1	1, 2, 3	Injection well	− 100
I2	1, 2, 3	Injection well	− 100
I3	1, 2, 3	Injection well	− 100
I4	1, 2, 3	Injection well	− 100

and production performance parameters of each well are shown in Table 4.2; the oil/water relative permeability curve is shown in Fig. 4.17.

The other parameters, porosity and effective thickness, take the same values in each layer, porosity is 0.2 and effective thickness is 5 m; permeability is homogeneous in each layer and heterogeneous among layers, horizontal

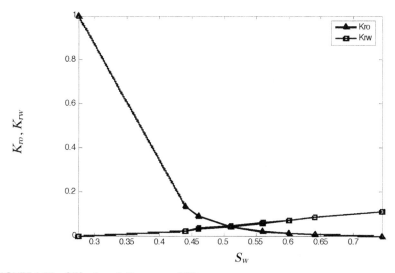

FIGURE 4.17 Oil/water relative permeability curve.

permeabilities of 1st, 2nd and 3rd layers (from the top down) are 0.2 μm², 0.3 μm² and 0.4 μm², respectively; cross-flow is not considered and vertical permeability is 0; oil and water viscosities are 9 mPa•s and 0.45 mPa•s, respectively; oil and water densities are 0.9 g/cm³ and 1 g/cm³, respectively.

Well P1 and I1 are taken as the testing wells, shut in the two wells and simulate pressure build-up and pressure drawdown test, respectively. The distribution of streamline and the oil saturation distribution along the streamlines in each layer of the two wells in the well shut period are shown in Figs 4.18 to 4.20. Variations in oil saturation in each streamline reflect the displacement effect of crude oil in each direction from the injection well. It can be seen that the third layer permeability is the largest and its displacement effect is the best; the oil saturation is the lowest after the same production period. In the region between injection wells of each layer (pressure equilibrium region), pressure gradient is low and fluid flows slowly, so the displacement effect is bad and oil saturation is high.

It can be seen from Fig. 4.21 that the pressure response of well P1 includes four periods. In the early and middle periods, no inter-layer interference and

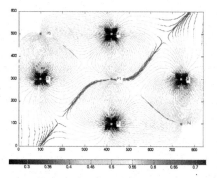

FIGURE 4.18   Oil saturation distribution along the streamlines in the first layer.

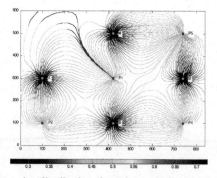

FIGURE 4.19   Oil saturation distribution along the streamlines in the second layer.

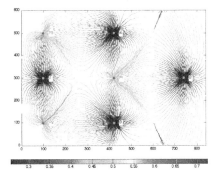

FIGURE 4.20   Oil saturation distribution along the streamlines in the third layer.

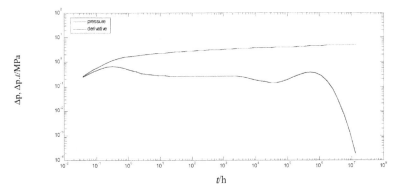

FIGURE 4.21   Pressure response curves of well P1.

boundary influence appear, pressure response is the same as that of a single-layer reservoir. In the middle and later periods, pressure derivative shows a concave, which is caused by the inter-layer interference created by multi-layer production. In the late period, pressure derivative falls, which shows the characteristic of constant pressure outer boundary. Because the production well is completely controlled by the surrounding injection wells, streamlines can not achieve the reservoir boundary, so this is the influence of the pseudo-constant pressure outer boundary formed by the injection wells.

It can be seen from Fig. 4.22 that pressure response of well I1 also includes four periods. In the early and middle periods, pressure response is the same as that of a single-layer reservoir. In the middle and later periods, pressure derivative shows upwarp due to the influence of the production wells. In the late period, due to the influence of reservoir constant pressure outer boundary, when pressure transmits to reservoir boundary, constant pressure difference is formed between well bore (inner boundary) and the outer boundary, "pseudo-steady flow" appears in the reservoir, pressure does not change with time and the pressure derivative falls.

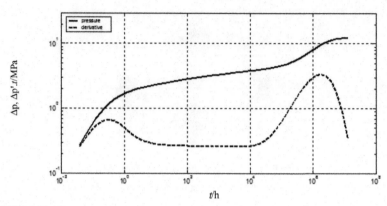

FIGURE 4.22   Pressure response curves of well I1.

4.4. LAYERING RATE RESPONSE CHARACTERISTICS OF THE STREAMLINE NUMERICAL WELL TESTING INTERPRETATION MODEL FOR A MULTI-LAYER SANDSTONE WATERFLOODING RESERVOIR

With the above streamline numerical well testing interpretation model for a multi-layer sandstone waterflooding reservoir, the underground rate of each layer can be calculated (as seen in Formula 4.1.17). In order to study the layering rate response characteristics of the streamline numerical well testing interpretation model for a multi-layer sandstone waterflooding reservoir, a model of a typical inverted five-spot pattern in a three-layer homogeneous reservoir is established, the main parameters are described below.

Porosity of each layer is 0.2, effective thickness is 50 m, and the permeabilities of each layer are $0.2 \times 10^{-3}\,\mu m^2$, $0.3 \times 10^{-3}\,\mu m^2$ and $0.5 \times 10^{-3}\,\mu m^2$, respectively. Oil and water phases take the same physical parameters and phase permeability, and the model is degenerated into a single-phase flow model. Fluid production of oil well P1~P4 is 20 m³/d, fluid injection rate of water well I is 100 m³/d. The 1st, 2nd and 3rd layers of well P1~P4 and I are all perforated.

Shut-in water injection well I at one time-point and simulate the pressure drawdown test: the following shows the layering rate results (Figs 4.23 to 4.25):

It can be seen from Fig. 4.23 that, in the radial flow period, the layering rate is stable and proportional to layer formation capacity (product of permeability and effective thickness). The pseudo-steady flow period begins after the pressure reaches that of neighboring wells (oil wells), since pressure difference is stable between the testing well (water well) and neighboring wells (oil wells), the layering rate will not change. Because the well pattern in this typical model is symmetrical, and the production rates of each production well are the same, although formation capacities of each layer are different, the final layering rates are equivalent.

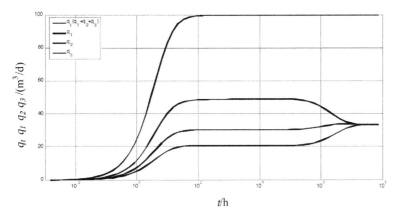

FIGURE 4.23   Layering rate curves with different layer formation capacity.

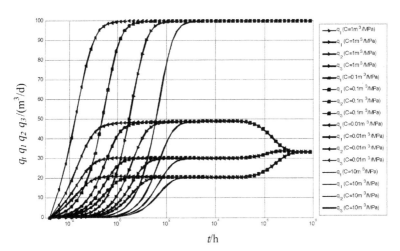

FIGURE 4.24   Layering rate curves with different wellbore storage factors.

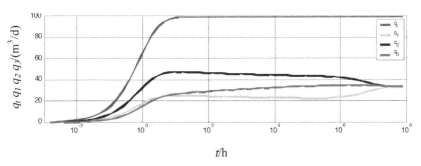

FIGURE 4.25   Layering rate with different layering skin factors.

Figure 4.24 shows layering rate curves with well bore storage factors taking $0.01\, m^3 / MPa$, $0.1\, m^3 / MPa$, $1\, m^3 / MPa$ and $10\, m^3 / MPa$, respectively. It can be seen that, the larger the well bore storage factor is, the later the time when layering and total rate achieve the biggest values. The pseudo-steady flow period begins after the pressure reaches that of neighboring wells (oil wells), pressure difference is stable between the testing well and neighboring wells, so the layering rates are equivalent and stable.

The above study shows the layering rate response characteristic when skin factors of each layer are the same. In order to further study the influence characteristic of layering skin factor on layering rate response in a multi-layer reservoir, the above model is modified as follows: skin factors of each layer are 0, 5 and 10, respectively; permeabilities of each layer are $0.1 \times 10^{-3}\,\mu m^2$, $0.2 \times 10^{-3}\,\mu m^2$ and $0.3 \times 10^{-3}\,\mu m^2$, respectively, the other parameters do not change.

Solve the typical mode and obtain the layering rate curves which are shown in Fig. 4.25. It can be seen that, because of the influence of the layering skin factor, the layering rate of each layer is not proportional to the layer formation capacity. Although the formation capacity of the third layer is the largest, its skin factor is the largest too, which causes the biggest additional pressure drop around the well bore, so its underground rate is small and even lower than the rate of the first layer which has the smallest formation capacity at the beginning. With the time passing, pressure transmits through the damage zone of the third layer, then the rate in the third layer rises slowly. The pseudo-steady flow period begins after the pressure reaches that of neighboring wells (oil wells), then the pressure difference is stable and equivalent between the testing well and neighboring wells, so the layering rates are stable and equivalent.

4.5. CHAPTER SUMMARY

In this chapter, based on the streamline numerical well testing interpretation model for a single-layer sandstone waterflooding reservoir, the numerical well testing interpretation model for a multi-layer sandstone waterflooding reservoir is studied, the main work includes:

1. The streamline numerical well testing interpretation model for a multi-layer sandstone waterflooding reservoir includes the filtration mathematical model of the production period and the streamline mathematical model of the testing period. The filtration mathematical model of the production period is a simplified multi-layer reservoir black oil model. The mathematical model of the testing period is the streamline model, which is made up of the simultaneous equations with the mathematical equation of each streamline in each layer from the testing well, and the equation along each streamline could consider the variation of multi-layer reservoir geology and fluid parameters. The pressure and saturation obtained from the filtration mathematical model of the production period are taken as the basement and

initial conditions to establish the streamline mathematical model of the testing period.

2. In the streamline numerical well testing interpretation model for a multi-layer sandstone waterflooding reservoir, the solving method of the filtration mathematical model in the production period is the same as that of single-layer sandstone waterflooding reservoir; i.e. use the IMPES method to solve pressure and use the streamline method to solve saturation. In order to solve the streamline mathematical model of the testing period, equations of all streamlines in all layers related to the testing well should be gathered together and solved.

3. From pressure drawdown simulation of the testing well in a multi-layer reservoir typical model, the influencing characteristics of well bore storage factor, skin factor and permeability ratio on pressure response and the pressure response characteristic with complex cross-flow and oil/water two-phase flow are studied.

   • With the increase of the well bore storage factor, the whole pressure response curve moves to the right and the appearance of radial flow delays; when inter-layer cross-flow exists, the influence of well bore storage effect on the beginning time of cross-flow is little, but too big well bore storage factor could cover the appearance of cross-flow.

   • When skin factors of each layer are equal, with different skin factors, pressure derivative curve in middle radial flow period coincides with that in later flow period; when skin factors of each layer aren't equal, with different skin factors, pressure derivative curve in radial flow period doesn't coincide with that in later flow period.

   • When cross-flow only exists from the low permeability layer to the high permeability layer, the pressure derivative curve shows as concave, the greater the permeability ratio is, the earlier the concave occurs and the deeper the concave will be.

   • When cross-flow exists from the high permeability layer to the low permeability layer, the pressure derivative curve may show upwarp.

4. The influencing characteristics of formation factor, well bore storage factor and skin factor on layering rate of the streamline numerical well testing interpretation model for a multi-layer sandstone waterflooding reservoir are studied. At present, layering reservoir testing technology is developing quickly; studying layering rate characteristics of multi-layer reservoirs and applying the layering rate data into automatic-fit interpretation of multi-parameter numerical well testing is important in reducing the ambiguity of multi-layer reservoir well testing interpretation and in obtaining the reliable layering characteristic parameter.

Streamline Numerical Well Testing Interpretation Model under Complex Near-Well-Bore Conditions

5.1. STREAMLINE NUMERICAL WELL TESTING INTERPRETATION MODEL OF A PARTIALLY PERFORATED WELL

5.1.1. Streamline Numerical Well Testing Interpretation Model of a Partially Perforated Well

The streamline numerical well testing interpretation model of a partially perforated well includes the filtration mathematical model of the production period and the streamline mathematical model of the testing period.

The filtration mathematical model of the production period of a partially perforated well is the same as that of a totally perforated well. However, the streamline mathematical model of the testing period of a partially perforated well is a little different from that of totally perforated well. The difference is that the node thickness of the streamline mathematical model in the testing period of a partially perforated well is not the whole effective thickness, but the production thickness, which is indicated with the vertical maximum swept thickness of streamlines in the model. The streamline mathematical model of a partially perforated well in the testing period is presented below.

5.1.1.1. Pressure Control Equation

Make subscript j represent the streamline order, and assume that the total number of streamlines from the testing well is N, then the flow control equation along each streamline is

$$\frac{1}{l_j}\frac{\partial}{\partial l_j}\left(l_j\frac{\lambda_{tj}h_{pj}}{\phi_j}\frac{\partial p_j}{\partial l_j}\right) = h_{pj}c_t\frac{\partial p_j}{\partial t} \ (j = 1, 2, \cdots, N) \tag{5.1.1}$$

where, h_{pj} is the production thickness crossed through by the jth streamline (cm).

Streamline Numerical Well Test Interpretation. DOI: 10.1016/B978-0-12-386027-9.00005-8
Copyright © 2011 by Elsevier Ltd

5.1.1.2. Inner Boundary Condition

The inner boundary condition includes the inner boundary condition for the production well and the injection well—two cases, which are shown in Equations 5.1.2 and 5.1.3, respectively.

$$\sum_{j=1}^{N} \frac{2\pi}{N} r_w h_{p1} \cdot \left(\lambda_{t1} \frac{\partial p_j}{\partial l_j} \right)_{l_j=r_w} = -q + C \left[\frac{dp_w}{dt} - Sl_j \frac{d}{dt} \left(\frac{\partial p_j}{\partial l_j} \right)_{l_j=r_w} \right] \quad (j=1,2,\cdots,N)$$

(5.1.2)

$$\sum_{j=1}^{N} \frac{2\pi}{N} r_w h_{p1} \cdot \left(\lambda_{t1} \frac{\partial p_j}{\partial l_j} \right)_{l_j=r_w} = q + C \left[\frac{dp_w}{dt} - Sl_j \frac{d}{dt} \left(\frac{\partial p_j}{\partial l_j} \right)_{l_j=r_w} \right] \quad (j=1,2,\cdots,N)$$

(5.1.3)

where, h_{pl} is the perforated thickness of the testing well.

5.1.1.3. Other Equations

Outer boundary conditions, initial conditions and the saturation equation are the same with the streamline numerical well testing interpretation model of a totally perforated well.

5.1.2. Pressure Response Characteristics of a Partially Perforated Well

5.1.2.1. Influencing Characteristics of Permeability Ratio on Pressure Response in the Well Testing Interpretation Model

In order to study the characteristics of the streamline numerical well testing interpretation model of a partially perforated well, a typical model of a single-layer homogeneous and uniform thickness reservoir with an inverted five-spot pattern is established, and the main parameters are described below.

Reservoir porosity is 0.2, formation thickness is 200 m, and real permeability is $1 \times 10^{-3} \, \mu m^2$. Well I is partially perforated in the middle zone, perforation thickness is 40 m; the other wells are completely perforated. Fluid production rate of oil well P1, P2 and P3 is 80 m³/d, fluid production rate of well P4 is 60 m³/d, injection rate of water well I is 100 m³/d. The perforated intervals of each well are shown in Fig. 5.1.

First, the influencing characteristics of permeability ratio (the ratio of vertical permeability to horizontal permeability) on the streamline numerical well testing interpretation model of a partially perforated well is studied.

The streamline distribution with permeability ratios 0.001, 0.05 and 1, respectively, are shown in Figs 5.2 to 5.4. It can be seen that, with different permeability ratios, the streamline patterns around the well bore of a partially perforated well (well I) are different; with the increase in permeability ratio, the included angle between well bore surrounding streamlines and vertical direction in a partially perforated well decreases; and different permeability ratios cause

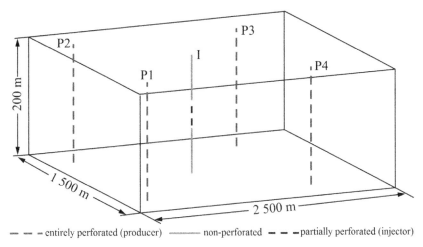

FIGURE 5.1   Schematic of perforated intervals in each well.

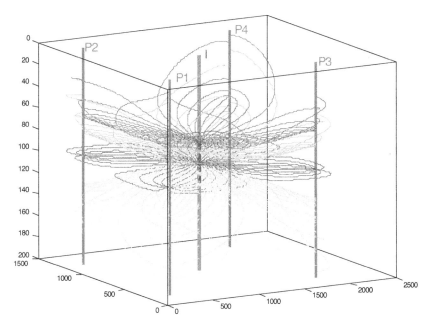

FIGURE 5.2   Streamline distribution with permeability ratio of 0.001.

different horizontal distances between streamlines and perforated well bore
when the vertical swept zone of the streamlines achieves the maximum production thickness (the whole formation thickness); the greater the permeability ratio
is, the shorter the horizontal distance between streamlines and perforated well
bore when the vertical swept zone of streamlines achieves the maximum production thickness will be.

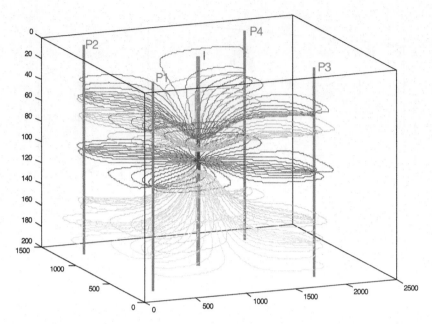

FIGURE 5.3   Streamline distribution with permeability ratio of 0.05.

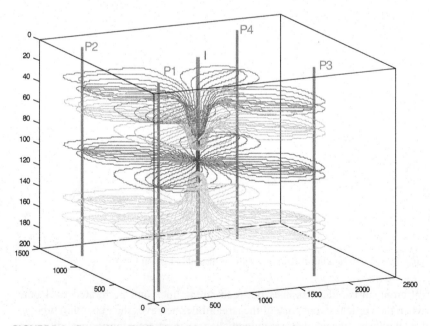

FIGURE 5.4   Streamline distribution with permeability ratio of 1.

Figure 5.5 shows the pressure response curves with three different permeability ratios; permeability ratios of the three pressure derivative curves are 0.001, 0.05 and 1 from top to bottom, respectively. In Fig. 5.5, pressure derivative

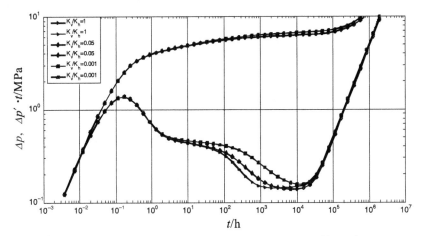

FIGURE 5.5   Pressure response contrast with three different permeability ratios.

curves transit from one horizontal straight segment (radial flow around well bore perforated thickness zone) to another horizontal straight segment (radial flow in the whole reservoir thickness zone far from the well bore), and the height of the second straight segment is lower than that of the first straight segment. This is the typical characteristic of spherical flow in a partially perforated well bore, which is caused by different production thickness. Around the well bore, the reservoir is partially perforated, the production thickness is only the thickness of the partially perforated zone; with the increase in distance from well bore, the vertical swept area of streamlines is greater, hence the pressure derivative curve falls (region between the two horizontal straight segments); when the vertical swept area of the streamlines achieves the maximum (the whole reservoir thickness), the fluid flow enters into another radial flow period, and the second horizontal straight segment appears on the pressure derivative curve.

From the contrast of streamline distributions in Figs 5.2 to 5.4, and pressure response curves in Fig. 5.5, we can see that, under the same reservoir conditions, if the vertical permeability is greater: (i) the lasting time of radial flow period around well bore is shorter; (ii) the starting time of cross-flow of streamlines in the vertical direction is earlier and the fall time of the pressure derivative curve is earlier; (iii) the horizontal distance between streamlines and partially perforated well bore when the vertical swept zone of streamlines achieves the maximum production thickness is shorter, and the beginning time of the radial flow period in the whole reservoir thickness is earlier.

5.1.2.2. Influencing Characteristics of Perforation Thickness on Pressure Response in the Well Testing Interpretation Model

Based on the above typical model, keeping the permeability ratio constant, the influencing characteristics of different perforated thicknesses on the pressure response of the streamline numerical well testing interpretation model of a partially perforated well is studied.

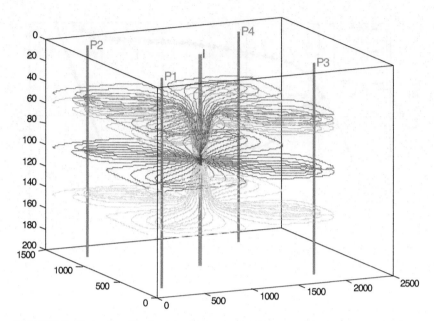

FIGURE 5.6 Streamline distribution with perforated thickness ratio of 1/10.

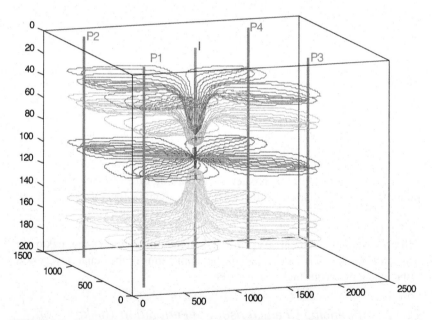

FIGURE 5.7 Streamline distribution with perforated thickness ratio of 1/5.

The streamline distributions with perforated thicknesses of 10 m and 20 m, respectively, are shown in Figs 5.6 and 5.7. It can be seen that, under certain formation and production conditions, different perforated thicknesses cause

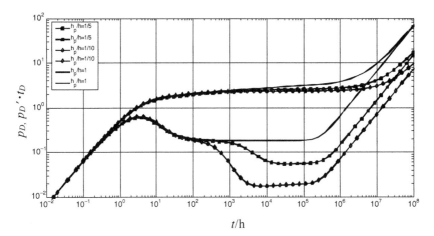

FIGURE 5.8   Pressure response contrast with different perforated thickness ratio.

different vertical maximum swept areas of streamlines. Obviously, the greater the perforated thickness is, the larger the vertical maximum swept area of streamlines will be.

Figure 5.8 shows dimensionless pressure response curves with different perforated thickness ratios. For contrast convenience, the thickness is dimensionless; definitions are by the perforated thickness instead, which means that

$$C_D = \frac{C}{2\pi\phi h_p c_t r_w^2}, \; t_D = \frac{K}{\phi\mu c_t r_w^2}t, \; \frac{2\pi K h_p}{q\mu B}\Delta p$$

It can be seen from Fig. 5.8 that the bigger the perforated thickness ratio, the smaller the height difference between two radial flow horizontal straight segments.

When the perforated thickness ratio is 1 (namely completely perforated), the two radial flow straight segments will combine into one horizontal straight segment. It can be seen from the above study that the second radial flow horizontal straight segment is caused by the pressure response of fluid flow in the whole reservoir area, so the greater the difference between perforated reservoir thickness and the whole reservoir thickness (the smaller the perforated thickness ratio) is, the greater the height difference between the two radial flow horizontal straight segments will be.

5.1.2.3. Influencing Characteristics of Perforation Position on Pressure Response in the Well Testing Interpretation Model

Based on the above typical model, the influencing characteristic of perforation position on the streamline numerical well testing interpretation model of a partially perforated well is studied.

Figures 5.9 and 5.10 present the streamline distributions with the perforation positions on the top and in the middle, respectively; Fig. 5.11 shows the pressure response with different perforation positions.

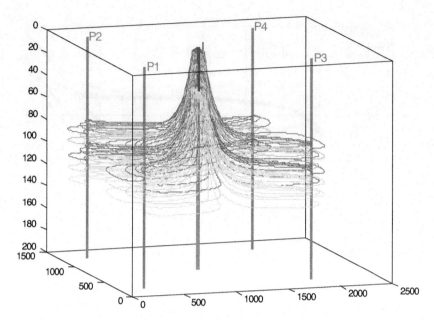

FIGURE 5.9   Streamline distribution with reservoir top perforated.

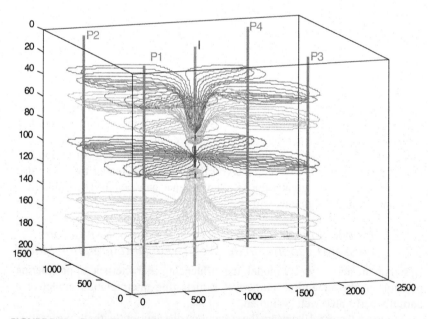

FIGURE 5.10   Streamline distribution with reservoir middle perforated.

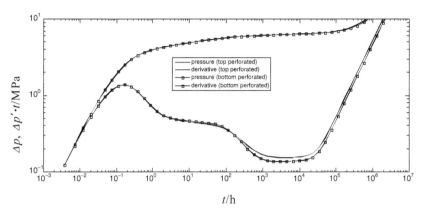

FIGURE 5.11   Pressure response contrast with different perforation position.

It can be seen from Fig. 5.11 that, under the production and formation conditions, there is not much difference in the pressure response curve pattern between top and middle perforated situations; only a little difference exists in the height of the second radial flow horizontal straight segment. The height of the second radial flow horizontal straight segment of the top perforated situation is greater than that of the middle perforated situation; it is not difficult to image that this is because, if the reservoir is top perforated, the maximum swept reservoir thickness area is less than that of the middle perforated situation.

5.2. STREAMLINE NUMERICAL WELL TESTING INTERPRETATION MODEL OF THE TESTING WELL WITH IRREGULARLY DAMAGED ZONE

5.2.1. Streamline Numerical Well Testing Interpretation Model of a Testing Well with Irregularly Damaged Zones

The streamline numerical well testing interpretation model of the testing well with irregularly damaged zones includes the filtration mathematical model of the production period and the streamline mathematical model of the testing period; the filtration mathematical model of the production period is the same as that of a testing well with regularly damaged zones, and the streamline mathematical model of the testing period is described as follows.

5.2.1.1. Pressure Control Equation

Make subscript j represent the streamline order, assume the total number of streamlines from the testing well is N, then the flow control equation along each streamline is

$$\frac{1}{l_j}\frac{\partial}{\partial l_j}\left(\alpha l_j \frac{\lambda_{tj}}{\phi_j}\frac{\partial p_j}{\partial l_j}\right) = \alpha c_t \frac{\partial p_j}{\partial t} \quad (j = 1, 2, \cdots, N) \qquad (5.2.1)$$

5.2.1.2. Inner Boundary Condition

Well bore storage effect and skin effect are considered. For well bore storage inner boundary, the testing well is separated into two situations including the production well and the injection well, and their inner boundary conditions are expressed in Equations 5.2.3 and 5.2.4, respectively:

$$\sum_{j=1}^{N} \frac{2\pi}{N} r_w h_1 \left(\lambda_{tj} \frac{\partial p_j}{\partial l_j} \right)_{l_j=r_w} = -q + C \frac{dp_{wf}}{dt} \quad (j=1,2,\cdots,N) \qquad (5.2.2)$$

$$\sum_{j=1}^{N} \frac{2\pi}{N} r_w h_1 \left(\lambda_{tj} \frac{\partial p_j}{\partial l_j} \right)_{l_j=r_w} = q + C \frac{dp_{wf}}{dt} \quad (j=1,2,\cdots,N) \qquad (5.2.3)$$

Skin inner boundary condition (simultaneous condition of well-bottom flow pressure):

$$p_{wf} = p_{wj} - S_j \left(l_j \frac{\partial p_j}{\partial l_j} \right)_{l_j=r_w} \quad (j=1,2,\cdots,N) \qquad (5.2.4)$$

From the Darcy formula, the rate of each streamline is calculated with the following formula:

$$q_j = u_t \cdot A = \lambda_{tj} \left(\frac{\partial p_j}{\partial l_j} \right)_{l_j=r_w} \cdot \left(\frac{2\pi}{N} r_w h_1 \right) \quad (j=1,2,\cdots,N) \qquad (5.2.5)$$

where, S_j is the skin factor of the jth streamline, u_t is the total velocity, and A is the area.

5.2.1.3. Other Equations

Outer boundary condition, initial condition and saturation equations are same with the streamline numerical well testing interpretation model with regularly damaged zones.

The solution of the streamline numerical well testing interpretation model with irregularly damaged zones is the same as that with regularly damaged zones (see details in Chapter 3).

5.2.2. Pressure Response Characteristics of a Well with Irregularly Damaged Zones

In order to study the pressure response characteristics of a streamline numerical well testing interpretation model with an irregularly damaged zone, a typical model of a nine-spot pattern in a single-layer homogeneous reservoir is established. In this model, effective thickness is 10 m, permeability is $0.02 \times 10^{-3} \, \mu m^2$, porosity is 0.2, fluid production rate of oil well P is $100 \, m^3/d$, injection rate of water well I1~I8 is $40 \, m^3/d$.

Take Oil well P as the testing well and simulate its pressure drawdown test. For research convenience, production history is not considered; select eight

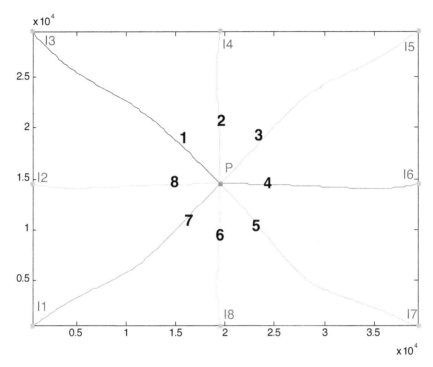

FIGURE 5.12 Streamline distribution diagram.

representative streamlines (the order is from 1 to 8 as seen in Fig. 5.12; the number of streamlines will not affect solution results in a homogeneous reservoir model) to establish the streamline mathematical model of the testing period, and then solve the model.

The pressure response curves when the skin factors of each streamline are equal are shown in Fig. 5.13. In this case, no matter how big the skin factor is, the damaged zone is circular, and the pressure derivative curves coincide into one horizontal line in radial flow period.

Figure 5.14 shows pressure response curve contrast of the testing well with regularly damaged zones (skin factor of each streamline is equal to 10) and irregularly damaged zones (skin factors of streamline 1 to 4 are 20, the others are 0). In this book, the well where all the skin factors of each streamline are not 0 is termed a "totally damaged" well, and the well where skin factors of partial streamlines are not 0 is termed a "locally damaged" well. It can be seen that the pressure derivative curves of locally and totally damaged wells do not coincide in the radial flow period.

Figure 5.15 shows the pressure response contrast of an undamaged well (skin factors of each streamline are 0), a locally damaged well (skin factor of streamline 1 is 10, the others are 0) and a totally damaged well (skin factors of each streamline are 10).

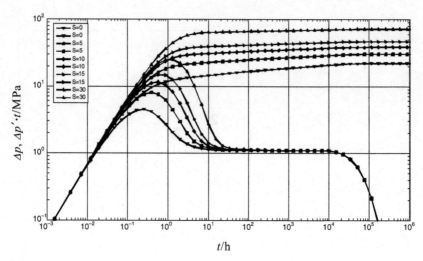

FIGURE 5.13   **Pressure response with regularly (circular) damaged zones.**

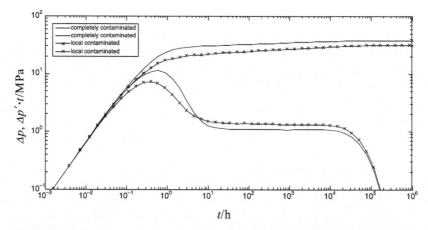

FIGURE 5.14   **Pressure response contrast between "locally damaged" and "totally damaged" wells.**

Skin factor S is used to represent the degree of damage; the well bore could be considered as being blocked off when S is very large. Here, well bore block percentage is described as the ratio of the streamline number with S equaling 10^{10} to the total streamline number (S of other streamlines are 0). Figure 5.16 shows the pressure response curve contrasted with different block percentages. It can be seen that, with the increase in the block percentage, pressure and derivative curves all move up, but the distance between pressure and derivative curves varies little.

It can be seen from the above simulation that the location of the pressure derivative in the radial flow period with irregularly damaged zones (including

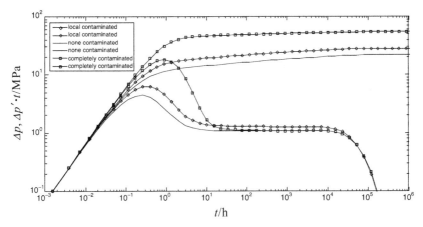

FIGURE 5.15   **Pressure response contrast of undamaged, locally damaged and totally damaged wells.**

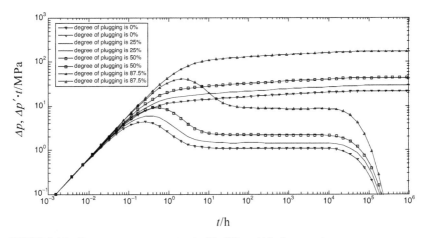

FIGURE 5.16   **Pressure response contrast with different block percentages.**

locally damaged well, locally blocked well and the well with different skin factors in each direction) is uncertain. It can also be seen from solving the streamline rate that the streamline rate of irregularly damaged zones has a close relationship with the damage degree of the near-well zone. Figure 5.17 shows the streamline rate response curves with different skin factors (obtained in formula 5.2.5), we can see the existence of skin factor produces a pressure drop loss (additional pressure drop) in the near-well region. In homogeneous formation conditions, with the increase of streamline skin factor, the pressure drop of the testing well in this direction increases, and the producing pressure drop decreases, so the streamline rate will decrease; different skin factors cause different pressure drop loss in the near-well region, and the pressure gradient

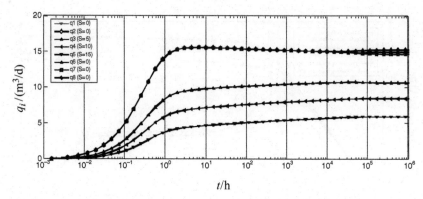

FIGURE 5.17 Rate response of the testing well with irregularly damaged zones.

distribution in the near-well region is irregular, which makes the uncertainty of the straight segment location on pressure derivative curve in the radial flow period.

5.3. CHAPTER SUMMARY

In this chapter, based on the streamline numerical well testing interpretation model with conventional inner boundary conditions (reservoir totally perforated, well bore storage effect and skin effect), the streamline numerical well testing interpretation model with complex near-well conditions is studied, the main work includes the points listed below.

5.3.1. Streamline Numerical Well Testing Interpretation Model of a Partially Perforated Well

The streamline numerical well testing interpretation model of a partially perforated well includes the filtration mathematical model of the production period and the streamline mathematical model of the testing period: the filtration mathematical model of the production period is the same with the totally perforated well; however, the streamline mathematical model of the testing period is different from that of the totally perforated well, the difference is that the streamline node thickness in the streamline mathematical model of the testing period is not the whole effective thickness but the production thickness (represented as the maximum swept thickness of streamlines in vertical direction).

With the pressure drawdown simulation of a partially perforated well in a typical model, the influencing characteristics of permeability ratio, perforation thickness and perforation position on the pressure response of the testing well are studied: (i) the pressure derivative curve shows two horizontal straight segments, which represent the radial flow of the near-well-bore perforated zone and the radial flow of the whole reservoir thickness region far from the well bore, and the second radial flow straight segment is lower than the first straight segment; (ii) in the same condition, the larger the vertical permeability is, the shorter the lasting time of the radial flow period around the well bore is, the earlier the cross- flow of streamline begins (the pressure derivative curve falls) in the vertical direction, and the earlier the radial flow in the whole reservoir thickness region occurs; (iii) in the same condition, the larger the reservoir perforated thickness ratio is, the smaller the height difference between the two radial flow straight segments will be, when perforated thickness ratio is 1 (totally perforated), the two radial flow straight segments coincide into one radial flow straight segment; (iv) in the same condition, the influence of perfo-ration location on the shape of the pressure response curve is small, there is only a small difference in the height of the horizontal straight segment in the second radial flow period. Generally, the height of the second horizontal straight seg-ment when the reservoir top is perforated is a little higher than that when the reservoir middle is perforated.

5.3.2. The Streamline Numerical Well Testing Interpretation Model of the Testing Well with Irregularly Damaged Zones

At present, all the commercial well testing interpretation software use only one skin factor to describe the well-bottom pollution, and the damaged zone is circular in this case. However, in streamline numerical well testing interpreta-tion model of the testing well with irregularly damaged zone, well testing equations could be established along each streamline, and different skin factors can be assigned to each equation, so the shape of the damaged zone is irregular.

From pressure build-up simulation of the testing well in a typical model, the pressure response characteristics of the testing well with irregularly damaged zones are studied:

- when damage to the testing well in each direction is the same, the pressure derivative curves with different skin factors will coincide in the radial flow period;
- when damage to the testing well in each direction is different, the pressure derivative curves with different skin factors will not coincide in the radial flow period.

Streamline Numerical Well Testing Interpretation Model for a Chemical Flooding Multi-Layer Sandstone Reservoir

6.1. STREAMLINE NUMERICAL WELL TESTING INTERPRETATION MODEL FOR A POLYMER FLOODING MULTI-LAYER SANDSTONE RESERVOIR

The streamline numerical well testing interpretation model for a polymer flooding multi-layer reservoir includes a filtration mathematical model of the production period and a streamline mathematical model of the testing period.

6.1.1. Filtration Mathematical Model of the Production Period
6.1.1.1. Basic Differential Equations

Based on the following assumptions:

1. Oil and water phases are involved in the fluid which contains three components. Oil phase contains oil component only, and water phase contains water and polymer components.
2. Reservoir rock and fluid are incompressible.
3. Generalized Darcy law is applicable to flow in porous media; generalized Fick law is applicable to dispersion.
4. Polymer solution decreases water phase permeability only.
5. The relationship between oil and water relative permeability doesn't change with the variation of components in water phase.
6. Influence of polymer on mass conservation in water solution is negligible.
7. Influence of gravity and capillary force are not considered.

The filtration mathematical model of the production period for a chemical flooding multi-layer reservoir is established as follows:

Oil component (oil phase) flow equation:

$$\nabla \cdot \left(\frac{\rho_o K K_{ro}}{\mu_o} \nabla p \right) + q_o = \frac{\partial(\phi \rho_o S_o)}{\partial t} \tag{6.1.1}$$

Water component (water phase) flow equation:

$$\nabla \cdot \left(\frac{\rho_w K K_{rw}}{R_K \mu_p} \nabla p \right) + q_w = \frac{\partial(\phi \rho_w S_w)}{\partial t} \tag{6.1.2}$$

Polymer component convective diffusion equation:

$$\begin{aligned} &\nabla \cdot \left(d_{pol} \phi S_w \nabla C_{pol} \right) - \nabla \left(C_{pol} u_w \right) + q_w C_{pol} \\ &= \frac{\partial\left(\phi S_w C_{pol} \right)}{\partial t} + \frac{\partial\left[F_{pol} \rho_R (1 - \phi) S_w \hat{C}_{pol} \right]}{\partial t} \end{aligned} \tag{6.1.3}$$

Saturation normalization equation:

$$S_o + S_w = 1 \tag{6.1.4}$$

where:

μ_{pol} is the viscosity of polymer solution (mPa•s);
R_K is the permeability decreasing coefficient (dimensionless);
d_{pol} is the diffusion coefficient of polymer component (cm^2/s);
C_{pol} is the mass concentration of polymer (g/g);
u_w is the water phase flow velocity (cm/s);
$\hat{C}_{pol}$ is the adsorbed mass fraction of polymer component (polymer mass adsorbed by unit mass oil sand) (g/g);
F_{pol} is the accessible pore volume fraction of polymer component (fraction);
R_p is the rock density (g/cm^3).

6.1.1.2. Physicochemical Parameter Equation
6.1.1.2.1. Viscosity of Polymer Solution
The polymer solution, whose viscosity is the function of shear rate, is a kind of non-Newtonian fluid, and viscosity is related to solution concentration, salinity and shearing. The relationship between polymer solution viscosity and shear rate is given by the Meter and Bird formula:

$$\mu_{pol} = \mu_w + \frac{\mu_0 - \mu_\infty}{1 + \left(\gamma/\gamma_{1/2} \right)^{p_a - 1}} \tag{6.1.5}$$

where:

μ_{pol} is the viscosity of polymer at some shear rate (mPa•s);
u_w is the water viscosity without polymer (mPa•s);
μ_0 is the viscosity of polymer solution at zero shear rate condition (mPa•s);
μ_∞ is the viscosity at infinite shear rate, approximated equal to water viscosity (mPa•s);
γ is the shear rate of polymer solution (s^{-1});

$\gamma_{1/2}$ is the shear rate with the polymer viscosity which is in the middle of zero shear rate viscosity and infinite shear rate viscosity, which is constant for a certain concentration (s^{-1});

p_α is the correlation factor between polymer solution viscosity and shear rate, which is constant for polymer with certain concentration, dimensionless.

Under fixed shear rate, the polymer solution viscosity is the function of polymer concentration and solution dielectric concentration, which is expressed in the following formula:

$$\mu_{pol} = \mu_w[1 + \left(A_{p1} C_{pol} + A_{p2} C_{pol}^2 + A_{p3} C_{pol}^3\right) C_{sep}^{s_p}] \qquad (6.1.6)$$

where:

A_{p1}, A_{p2}, A_{p3} is the viscosity parameter of polymer solution (g/cm^3);
s_p is the correlation factor of salt effect in polymer solution (dimensionless);
c_{sep} is the dielectric concentration.

6.1.1.2.2. Polymer Adsorption

Polymer adsorption is considered easily. Here, chemical materials adsorb on the rock surface while flowing through the porous media, the adsorption is irreversible and dynamic balance is established instantaneously, adsorption regularly follows the Langmuir adsorption isotherm:

$$\hat{C}_{pol} = \frac{aC_{pol}}{1 + bC_{pol}} \qquad (6.1.7)$$

where a and b are factors determined by experiments (dimensionless).

6.1.1.2.3. Permeability Decreasing Coefficient

Permeability decreasing coefficient is defined as follows:

$$R_K = \frac{K_w}{K_{pol}} \qquad (6.1.8)$$

where:

K_w is the water permeability (μm^2);
K_{pol} is the polymer solution permeability (μm^2).

Permeability decreasing coefficient is calculated using the following equation:

$$R_K = 1 + \frac{(R_{Kmax} - 1) \cdot C_K \cdot C_{pol}}{1 + C_K \cdot C_{pol}} \qquad (6.1.9)$$

where:

R_{Kmax} is the maximum permeability decreasing coefficient measured by experiments, which is constant for fixed polymer (dimensionless);
C_K is the constant measured by experiments (dimensionless).

6.1.1.2.4. Resistance Factor and Residual Resistance Factor

Resistance factor is the ratio of water mobility and polymer solution mobility while flowing through porous media, which is expressed as follows:

$$R_F = \frac{K_w}{K_{pol}} \frac{\mu_{pol}}{\mu_w} = R_K \frac{\mu_{pol}}{\mu_w} \qquad (6.1.10)$$

Residual resistance factor is the ratio of water mobility between polymer solution flow through porous media before and after, which is mainly caused by the polymer adsorption. This model is calculated in the following formula:

$$R_{RF} = 1 + \sum l_i ads_i \qquad (6.1.11)$$

where ads_i is the effective adsorption concentration of each material (g / g).

6.1.1.2.5. Inaccessible Pore Volume Fraction for Polymer

Because polymer can not achieve the pore space which is smaller than polymer, this kind of pore space is called inaccessible pore volume. Polymer can only enter the pore with bigger space. The main pore space that polymer can not enter is as follows: pore volume left due to the small pore radius which the polymer could not enter, pore volume blocked by polymer, and hydrodynamics volume by adsorption on rock surface.

The inaccessible pore volume is influenced by the factors including polymer molecular structure, pore size, pore and throat radius, permeability, fluid rheology, and salt concentration. Generally speaking, the greater the permeability is, the smaller the inaccessible pore volume is. However, permeability only can not be used to determine inaccessible pore volume, which can be approximately described as follows:

$$IPV = (\phi - \phi_{pol})/\phi \qquad (6.1.12)$$

where:

IPV is the inaccessible pore volume fraction (fraction);

ϕ_{pol} is the polymer accessible porosity (fraction).

6.1.2. Streamline Mathematical Model of the Testing Period

The streamline mathematical model for a polymer flooding multi-layer reservoir of the testing period is similar to that for water flooding. The difference is that the influence of saturation and polymer concentration should be considered when calculating mobility parameters of streamline nodes. The model can be established as described below.

6.1.2.1. Pressure Control Equation

Superscript m indicates layer number, subscript j streamline number, and the streamline numbers emitted from the testing well in layer m assumed as N^m. The flow control equation along each streamline can be described as follows:

$$\frac{1}{l_j^m}\frac{\partial}{\partial l_j^m}\left(\alpha l_j^m \frac{\lambda_{tj}^m}{\phi_m^j}\frac{\partial p_j^m}{\partial l_j^m}\right) = \alpha c_t \frac{\partial p_j^m}{\partial t} \quad (m=1,2,\Lambda,M)\ (j=1,2,\Lambda,N^m)$$

(6.1.13)

6.1.2.2. Inner Boundary Condition

Well bore storage effect and skin effect is considered. For well bore storage inner boundary, the testing well is divided into production well and injection well; the equations are 6.1.14 and 6.1.15, respectively.

$$\sum_{m=1}^{M}\sum_{j=1}^{N^m}\frac{2\pi}{N^m}r_w h_1^m \left(\lambda_{tj}^m\frac{\partial p_j^m}{\partial l_j^m}\right)_{l_j^m=r_w} = -q + C\frac{dp_{wf}}{dt}(m=1,2,\Lambda,M)\ (j=1,2,\Lambda,N^m)$$

(6.1.14)

$$\sum_{m=1}^{M}\sum_{j=1}^{N^m}\frac{2\pi}{N^m}r_w h_1^m \left(\lambda_{tj}^m\frac{\partial p_j^m}{\partial l_j^m}\right)_{l_j^m=r_w} = q + C\frac{dp_{wf}}{dt}(m=1,2,\Lambda,M)\ (j=1,2,\Lambda,N^m)$$

(6.1.15)

Skin inner boundary condition (bottom-hole flowing pressure simultaneous condition):

$$p_{wf}=p_{wj}^m - S^m \left(l_j^m\frac{\partial p_j^m}{\partial l_j^m}\right)_{l_j^m=r_w} \quad (m=1,2,\Lambda,M)\ (j=1,2,\Lambda,N^m)$$ (6.1.16)

6.1.2.3. Outer Boundary Condition

This is the same as the streamline mathematical model for a water flooding multi-layer reservoir of the testing period.

6.1.2.4. Initial Conditions

Initial conditions include the distribution of pressure, saturation, polymer concentration and streamline, which can be obtained by solving the filtration mathematical model for a polymer flooding multi-layer sandstone reservoir of the production period.

6.1.2.5. Saturation Equation

$$\frac{\partial S_w}{\partial t} + \frac{\partial f_{wp}}{\partial \tau} = 0$$ (6.1.17)

where f_{wp} is the water phase fractional flow (fraction). The definition and derivation of Formula 6.1.17 is explained in detail later.

6.1.2.6. Polymer Concentration Equation

$$\frac{\partial\left[F_{pol}\rho_R(\frac{1-\phi}{\phi})S_w\hat{C}_{pol}\right]}{\partial t} + \frac{\partial(S_w C_{pol})}{\partial t} + \frac{\partial(C_{pol}f_{wp})}{\partial\tau} = 0 \qquad (6.1.18)$$

The integrated streamline mathematical model for apolymer flooding multi-layer reservoir of the testing period can be shown from Equations 6.1.13 to 6.1.18.

6.2. SOLVING METHODS OF THE STREAMLINE NUMERICAL WELL TESTING INTERPRETATION MODEL FOR A POLYMER FLOODING MULTI-LAYER SANDSTONE RESERVOIR

6.2.1. Solving Methods of the Filtration Mathematical Model in the Production Period

6.2.1.1. Derivation of Saturation and Polymer Concentration Equation Along Streamline

The streamline method is used to solve saturation and polymer concentration in the filtration mathematical model of the production period. At first, convert two-dimensional or three-dimensional models based on network to a one-dimensional model, and the conversion method of saturation equation is the same with that of water flooding. The equation is described as follows:

$$\frac{\partial S_w}{\partial t} + \frac{\partial f_{wp}}{\partial\tau} = 0 \qquad (6.2.1)$$

where

$$f_{wp} = \frac{\lambda_{pol}}{\lambda_{pol} + \lambda_o} = \frac{\lambda_{pol}}{\lambda_t} \qquad (6.2.2)$$

$$\lambda_o = \frac{KK_{ro}}{\mu_o} \qquad (6.2.3)$$

$$\lambda_{pol} = \frac{KK_{rw}}{R_K\mu_{pol}} \qquad (6.2.4)$$

where f_{wp} is the water fractional flow (fraction).

Initial conditions and inner boundary conditions, respectively:

$$S_w(s,0) = S_{wi} \qquad (s \geq 0) \qquad (6.2.5)$$
$$S_w(0,t) = S_{w,inj} \qquad (t > 0) \qquad (6.2.6)$$

Then, neglecting dispersion and lateral diffusion, polymer concentration equation along the streamline is:

$$\frac{\partial\left[F_{pol}\rho_R(\frac{1-\phi}{\phi})S_w\hat{C}_{pol}\right]}{\partial t} + \frac{\partial(S_w C_{pol})}{\partial t} + \frac{\partial(C_{pol}f_{wp})}{\partial\tau} = 0 \qquad (6.2.7)$$

Initial conditions and inner boundary conditions, respectively:

$$C_{pol}(S, 0) = C_{pol,i} \qquad (S \geq 0) \qquad (6.2.8)$$

$$C_{pol}(0, t) = C_{pol,inj} \qquad (t > 0) \qquad (6.2.9)$$

where:

$C_{pol,i}$ is the initial polymer concentration of streamlines, which could be solved from polymer flooding filtration mathematical model of production period (g/g);

$C_{pol,inj}$ is the polymer concentration of injection well point (g/g).

6.2.1.2. Solution of Saturation and Polymer Concentration Equation Along Streamline

For polymer concentration Equation 6.2.7, first-order forward difference is used on time, and first-order central difference is used on space; the difference equation is obtained as follows:

$$\frac{(s_w C_{pol})_j^{n+1} - (S_w C_{pol})_j^{n}}{\Delta t} = -\frac{\left(C_{pol} f_{wp}\right)_{j+\frac{1}{2}}^{n} - \left(C_{pol} f_{wp}\right)_{j-\frac{1}{2}}^{n}}{\Delta \tau}$$
$$-\frac{\left[F_{pol}\rho_R \left(\frac{1-\phi}{\phi}\right) S_w\right]_j^{n+1} - \left[F_{pol}\rho_R \left(\frac{1-\phi}{\phi}\right) S_w\right]_j^{n}}{\Delta t} \qquad (6.2.10)$$

where:

C_{pol} and C_{pol} are the polymer concentrations of node j at time n and $n+1$, respectively (g/g);

$C_{pol j \pm \frac{1}{2}}^{n}$ is the polymer concentration of node $j \pm \{\frac{1}{2}\}$ at time n (g/g).

In order to solve the polymer concentration, first Equations 6.2.1, 6.2.5 and 6.2.6 are explicitly solved along the streamline to obtain the saturation; then the saturation is taken into Equations 6.2.7, 6.2.8 and 6.2.9, and the difference solution is used to get the polymer concentration.

6.2.2. Solving Methods of the Streamline Mathematical Model in the Testing Period

The solving methods of the streamline mathematical model for a polymer flooding multi-layer reservoir during the testing period are similar to those for water flooding, i.e. combine all the streamlines of all layers that are relevant to the testing well and solve together.

The difference is that the change of saturation and polymer concentration should be considered while updating mobility parameters in the streamline mathematical model for a polymer flooding multi-layer reservoir of the testing period. At first, use saturation Equation 6.2.1 to obtain saturations along streamline. Then, take the new saturation into the polymer concentration

Equation 6.2.7 and obtain the polymer concentration. Next, take the new polymer concentration into the polymer physicochemical parameter equation and calculate the total mobility of streamline nodes with the saturation parameter. Finally, the updated mobility parameter is obtained for the calculation of next time-step.

6.3. PRESSURE RESPONSE CHARACTERISTICS OF THE STREAMLINE NUMERICAL WELL TESTING INTERPRETATION MODEL FOR A POLYMER FLOODING MULTI-LAYER SANDSTONE RESERVOIR

6.3.1. Pressure Response Characteristics of the Well Testing Interpretation Model for a Polymer Flooding Single-Layer Reservoir

6.3.1.1. Influence of Initial Polymer Injection Concentration on Pressure Response

The following figure shows the characteristic curve of well testing pressure response in inverted five-spot production with polymer flooding, well bore storage factor $C = 0.05\,\text{m}^3/\text{MPa}$, skin factor $S = 0.001$ and take production well drawdown test as the calculation example. Basic reservoir parameters are in the following: reservoir area is 630 m × 630 m, layer thickness is 5 m, porosity is $0.8\,\mu\text{m}^2$, initial formation pressure is 20 MPa, and constant flow rate is $40\,\text{m}^3/\text{d}$ before testing well shut-in. Figure 6.1 shows pressure and its derivative curve with polymer concentrations of 500 mg/l, 1250 mg/l and 2000 mg/l in the system, respectively.

It is evident that there is an obvious depression in the pressure derivative curve of Fig. 6.1. The decrease in pressure derivative is influenced by the pseudo-constant pressure boundary of the surrounding injection well, then the outer closed boundary influences the testing well and causes the rise in

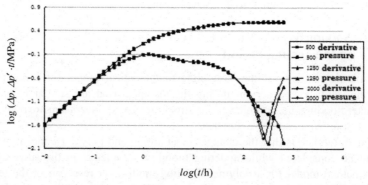

FIGURE 6.1   Pressure and its derivative curve with different polymer initial injection concentrations.

the derivative curve. With the increase of polymer injection concentration, the depression in the pressure derivative curve becomes earlier, which is caused by the smaller oil/water viscosity ratio after polymer injection. With the increase in injection concentration, the pseudo-constant pressure boundary of surrounding water wells is closer to the production well, and the depression of derivative curve will be earlier.

6.3.1.2. Influence Characteristic of Water Injection Time on Pressure Response After Polymer Flooding

Figure 6.2 shows the characteristic curve of well testing pressure response in a polymer flooding and inverted five-spot system, in which production wells are distributed at the four corners of the closed region and the injection well is located at the center of this region. Well bore storage factor $C = 0.05\,\text{m}^3/\text{MPa}$, skin factor $S = 0.001$, and take injection well drawdown test as the calculation example, basic parameters are same as the above. Figure 6.2 shows the pressure and its derivative curve, with continuously injected polymer for 30 days with polymer initial concentration of $1500\,\text{mg}/\text{l}$, and then injected with water at 30, 60, 90 and 120 days, respectively.

As seen in Fig. 6.2, the difference between the water well drawdown testing curve of polymer flooding and the typical injection well testing curve of single-phase flow is that the pressure derivative curve has an obvious rising tendency during the polymer slug region (with time log cycle 0–1). This is because the rise in polymer solution viscosity increases water viscosity obviously, and then decreases oil/water viscosity ratio, which causes a bigger pressure difference and the rising tendency of the pressure derivative curve. When it moves close to the oil/water front, the pressure derivative curve achieves the highest point, which is caused by pseudo-constant pressure boundary of the water well. Finally, the rising tendency of the pressure derivative curve is affected by the outer closed boundary.

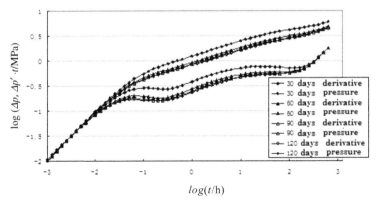

FIGURE 6.2   Pressure and its derivative curve with different water flooding time after polymer injection.

As shown by the different water flooding times after polymer flooding, the derivative curve does not change obviously with the increase in water flooding time. This is because polymer moves forward around the oil well and polymer concentration decreases around the water well.

6.3.1.3. Pressure Response Characteristics with High Permeable Banded Zone Distribution

6.3.1.3.1. High Permeable Banded Distribution Along the Main Streamline Direction

Figure 6.3 shows the characteristic curve of well testing pressure response in a polymer flooding inverted five-spot system, production wells distributed at the four corners of the closed region and injection well located at the center of this region, high permeable bands distribute along the main streamline direction of the diagonal oil/water wells. Well bore storage factor $C = 0.05\,m^3/MPa$, skin factor $S = 0.001$, take injection well drawdown test as the calculation example, and the basic parameters are same as the above. Figure 6.3 shows the pressure and its derivative curve, with continuously injected polymer for 30 days with polymer initial concentration of $1500\,mg/l$, and then injected with water.

As seen in Fig. 6.3, the difference between the water well drawdown testing curve of polymer flooding and the typical injection well testing curve of single-phase flow is that the pressure derivative curve has an obvious rising tendency during the polymer slug region.This is because the rise in polymer solution viscosity increases water viscosity obviously, and then decreases the oil/water viscosity ratio, which causes a bigger pressure difference and the rising tendency of the derivative curve. However, the rising tendency of the later derivative curve is due to the polymer cross-flow along high permeable bands of the water/oil well main streamline direction, and the influence of the pseudo-constant pressure boundary disappears in the oil/water front. The final rising tendency is caused by the outer closed boundary.

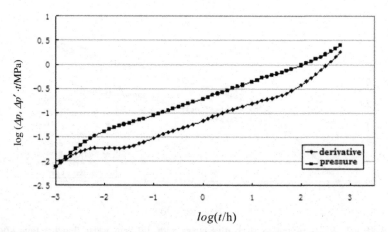

FIGURE 6.3   Pressure and its derivative curve of high permeable bands along main streamline direction.

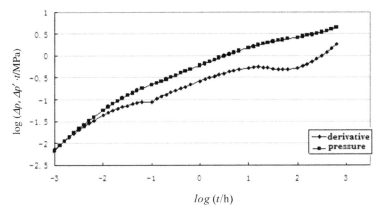

FIGURE 6.4   Pressure and its derivative curve of high permeable bands with shape S.

6.3.1.3.2. High Permeable Banded Distribution with S-Shape

Figure 6.4 shows the characteristic curve of the well testing pressure response in a polymer flooding inverted five-spot system, production wells distributed at the four corners of the closed region and injection well located at the center of this region, high permeable bands with S-shape distributed along the main streamline direction of the diagonal oil/water wells. Well bore storage factor $C = 0.05\,\text{m}^3/\text{MPa}$, skin factor $S = 0.001$, take injection well drawdown test as the calculation example, and the basic parameters are the same as the above. Figure 6.4 shows the pressure and its derivative curve, which continuously injected polymer 30 days with polymer initial concentration 1500 mg / l, and then inject water.

As seen in Fig. 6.4, the momentary horizon tendency disappears after the first hump on the derivative curve; this is also due to the influence of high permeable bands around injection wells. The later difference between the water well drawdown testing curve of polymer flooding and the typical injection well testing curve of single-phase flow is that the derivative curve has an obvious rising tendency during the polymer slug region. This is because the rise in polymer solution viscosity increases water viscosity obviously, and then decreases the oil / water viscosity ratio, which causes a bigger pressure differ-ence and the rising tendency of derivative curve. Because the influence of high permeable bands with S-shape is not as obvious as that along the oil / water well main streamline direction, the pressure derivative curve achieves the highest point, which is due to the influence of water well pseudo-constant pressure boundary. The final rising tendency is caused by the outer closed boundary.

6.3.1.3.3. High Permeable Banded Distribution Perpendicular to the Main Streamline Direction

Figure 6.5 shows the characteristic curve of well testing pressure response in a polymer flooding inverted five-spot system, production wells distributed at the four corners of the closed region and injection well located at the center of this region, high permeable bands are perpendicular to the main streamline direction

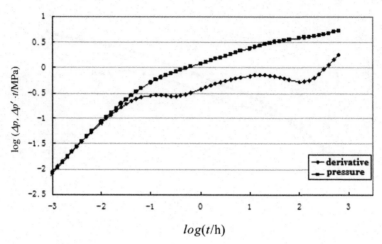

FIGURE 6.5 Pressure and derivative curve of high permeable bands perpendicular to main streamline direction.

of the diagonal oil / water wells. Well bore storage factor $C = 0.05 \text{ m}^3 / \text{MPa}$, skin factor $S = 0.001$, take injection well drawdown test as the calculation example, and the basic parameters are the same as the above. Figure 6.5 shows the pressure and its derivative curve, with continuously injected polymer for 30 days with polymer initial concentration of 1500 mg / l, and then inject water.

As seen in Fig. 6.5, the derivative curve shows high permeable bands perpendicular to the oil/water well main streamline direction, which does not affect the pressure measurement of injection well, and the shape of the derivative curve is similar, in that high permeable bands are absent.

At this time, the difference between the water well drawdown testing curve of the polymer flooding and typical injection well testing curve of single-phase flow is that the pressure derivative curve has an obvious rising tendency during the polymer slug region (with time log cycle 0–1). This is because the rise in polymer solution viscosity increases water viscosity obviously, and then decreases the oil/water viscosity ratio, which causes a bigger pressure difference and the rising tendency of derivative curve. When it moves close to the oil/water front, the pressure derivative curve achieves the highest point, which is the influence of the water well pseudo-constant pressure boundary. The final rising tendency of the pressure derivative curve is affected by the outer closed boundary.

6.3.2. Pressure Response Characteristics of the Well Testing Interpretation Model for a Polymer Flooding Multi-Layer Reservoir

6.3.2.1. Pressure Response Characteristics of Wells During the Process of Polymer Injection

In order to study pressure response characteristics of testing wells during the process of polymer injection for a polymer flooding multi-layer reservoir, a

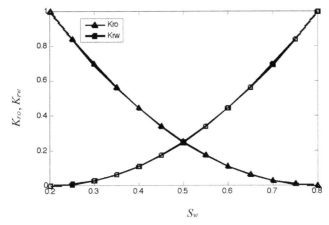

FIGURE 6.6   Oil/water relative permeability curve.

typical model of inverted nine-spot well pattern is built for two layered polymer flooding reservoirs, and the main parameters are as follows:

Different permeability values are assumed for each reservoir layer, with first (upper) layer permeability 0.8 μm², second (lower) layer permeability 2.4 μm² and vertical permeability 0; porosity and effective thickness are same for each layer with porosity 0.2 and effective thickness 2 m.

The viscosities of oil and water are 9 mPa•s and 0.45mPa•s, respectively; the densities of oil and water are 0.9 g/cm³ and 1 g/cm³, respectively; the production rate of oil well P1~P8 is 50 m³/d; the injection rate of polymer injection well I is 400 m³/d. The two layers of all oil and water wells are entirely perforated. The oil/water relative permeability curve is shown in Fig. 6.6.

After polymer flooding for 500 days under the same working system, the polymer injection well I is closed and pressure drawdown is simulated; the calculation results are shown as follows:

The oil saturation and polymer concentration distribution on streamlines after polymer flooding for 500 days are shown in Figs 6.7 to 6.10. It can be seen that the reservoir fluids distribute with the following characteristics: two circular distribution regions are formed around testing well I, namely the polymer slug region and the oil region, and the polymer slug region is nearer to the testing well; for the different permeability of each layer, the area of the polymer slug region on each layer is different. While permeability of the second layer is larger, so is the radius of the polymer slug region.

As seen in Fig. 6.11, the pressure derivative curve declines after the radial flow period; this is because total mobility increases when pressure transmits from the polymer slug region to the oil region, so the polymer flooding front could be judged using this method curve (the lowest point on pressure derivation).

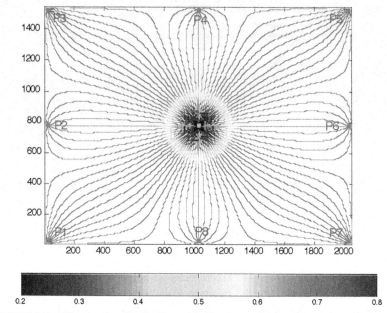

FIGURE 6.7 Oil saturation distribution on the first layered streamline.

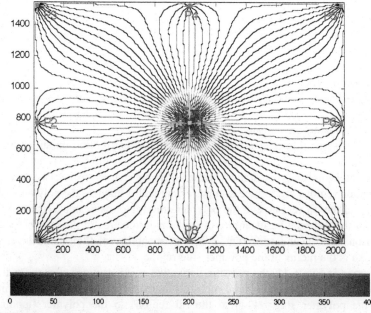

FIGURE 6.8 Polymer concentration distribution on the first layered streamline.

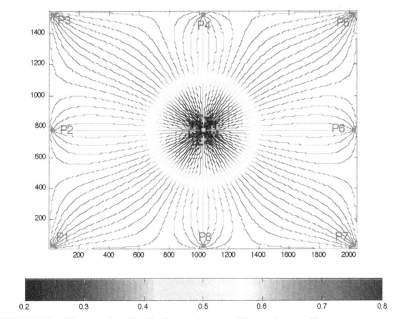

FIGURE 6.9   Oil saturation distribution on the second layered streamline.

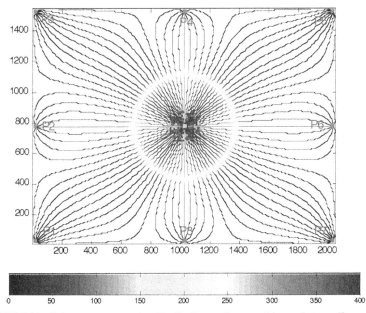

FIGURE 6.10   Polymer concentration distribution on the second layered streamline.

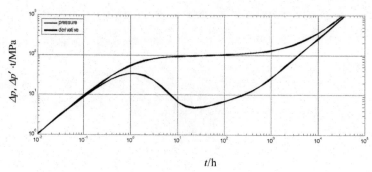

FIGURE 6.11   Pressure response of testing well.

6.3.2.2. Pressure Response Characteristics of Water Flooding Wells After Polymer Injection

As above, firstly with polymer flooding for 500 days under the same working system, then change to water flooding for 300 days; lastly close injection well I and simulate pressure drawdown; the calculation results are as follows.

Figures 6.12 to Fig. 6.15 show the oil saturation and polymer concentration distributions on streamlines after 500 days of polymer flooding followed by 300 days of water flooding. It can be seen that the reservoir fluids distribute with the following characteristics: three circular distribution regions are formed around testing well I, which are water region, polymer slug region and oil region, respectively; and the polymer slug region is located between the water

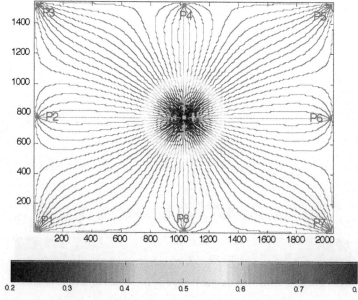

FIGURE 6.12   Oil saturation distribution on the first layered streamline.

FIGURE 6.13   Polymer concentration distribution on the first layered streamline.

and oil regions. Because of the different permeability of each layer, the area of water and polymer slug regions in each layer is different. While permeability of the second layer is larger, so is the radius of the water and polymer slug regions.

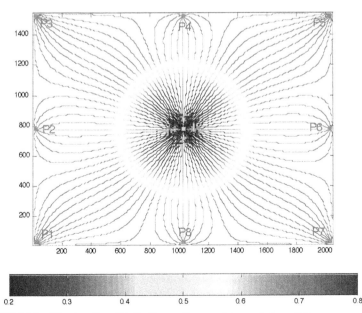

FIGURE 6.14   Oil saturation distribution on the second layered streamline.

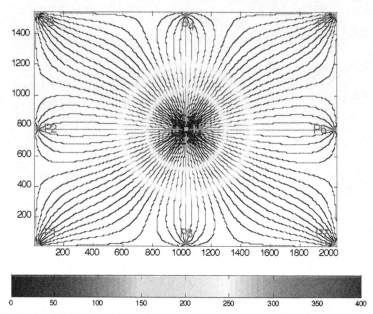

FIGURE 6.15   Polymer concentration distribution on the second layered streamline.

Among the water, polymer slug and oil regions, the viscosity of the polymer solution is the largest and the viscosity of the oil region is the second largest. While taking the testing well as the center and moving far away gradually, the mobility decreases at first (from the water region to the polymer slug region) and then increases (from the polymer slug region to the oil region).

As seen in Fig. 6.16, there are two "rises" and one "sink" in the pressure derivative curve. The first rise happens in the middle period and hides the radial flow straight segment; at this time, the pressure response transmits from the water region to the polymer slug region, the mobility decreases and fluid flow is

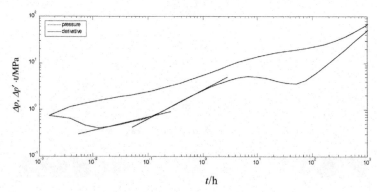

FIGURE 6.16   Pressure response after the testing well was closed.

dragged, then "rise" occurs in the pressure derivative curve; because of the different permeability and fluid distribution between the two layers in this period, the pressure response transmits to the polymer slug region over a different time, and the slope of pressure derivative curve changes (increases) constantly. The second "rise" in the pressure derivative is influenced by the well testing outer boundary. There is a "sink" in the derivative curve after two "rises": the pressure response transmits from the polymer slug region to the oil region, the mobility increases and the fluid flow resistance decreases. Because the mobility of the oil region is lower than that of the water region, the pressure derivation does not return to its former height but gets higher.

6.4. STREAMLINE NUMERICAL WELL TESTING INTERPRETATION MODEL FOR AN ALKALINE/POLYMER COMBINATION FLOODING MULTI-LAYER SANDSTONE RESERVOIR AND SOLVING METHODS

6.4.1. Filtration Mathematical Model and Resolution of Production Period

6.4.1.1. Basic Differential Equation

The basic differential equation for alkaline/polymer combination flooding of the production period is built with the following assumptions:

1. Reservoir is isothermal and phase equilibrium builds instantly.
2. Rock and fluids are incompressible; multi-phase flow obeys generalized Darcy law and dispersion follows generalized Fick law.
3. Oil and water phases are involved in the fluids which contain four pseudo-components. The oil is black oil and only contains oil component; water phase contains water, alkaline and polymer components; and there is no chemical reaction between the components. Chemical loss phenomena are taken into account, such as adsorption, chemical degradation, ion exchange and dissolution reaction, etc.
4. The existence of vapor and microemulsion phase is negligible, and the effect of alkaline and polymer on mass conservation in water solution is negligible.
5. The relationship between oil and water relative permeability does not change with the variety of components in water phase.
6. The effects of gravity and capillary force are negligible.
7. The influence of ion exchange and dissolution reaction on porosity and permeability is negligible.

Oil component (oil phase) flow equation:

$$\nabla \cdot \left[\frac{KK_{ro}}{B_o\mu_o} \nabla (p_o - \rho_o gD) \right] + q_o = \frac{\partial}{\partial t} \left(\frac{\phi S_o}{B_o} \right) \tag{6.4.1}$$

Water component (water phase) flow equation:

$$\nabla \cdot \left[\frac{KK_{rw}}{B_w R_K \mu_{wp}} \nabla(p_w - \rho_w gD) \right] + q_w = \frac{\partial}{\partial t} \left(\frac{\phi S_w}{B_w} \right) \qquad (6.4.2)$$

Alkaline component mass-transfer diffusion equation:

$$\nabla \cdot (d_{OH} \phi S_w \nabla c_{OH}) - \nabla(c_{OH} v_w) - \epsilon_{OH} c_{OH}$$
$$= \frac{\partial(\phi S_w c_{OH})}{\partial t} + \frac{K_a}{(1 + K_b c_{OH})^2} \frac{\partial[(1 - \phi) S_w c_{OH}]}{\partial t} \qquad (6.4.3)$$

Polymer component convective diffusion equation:

$$\nabla \cdot (d_p \phi S_w \nabla c_p) - \nabla(c_p v_w) + \phi S_w R_p + q_w \rho_w c_p$$
$$= \frac{\partial(\phi S_w c_p)}{\partial t} + \frac{\partial[F_p \rho_R (1 - \phi) S_w \widehat{c}_p]}{\partial t} \qquad (6.4.4)$$

where:

d_{OH} is the dispersion coefficient of alkaline component (cm^2/s);
c_{OH} is the mass concentration of injected alkali (g/g);
ϵ_{OH} is the alkaline chemical reaction coefficient (s^{-1});
K_a is the parameter that describes ion exchange and adsorptive capacity (fraction);
K_b is the alkaline adsorption constant (g/g);
v_w is the water phase flow velocity (cm/s).

Adding boundary conditions, initial conditions and the alkaline/polymer physicochemical parameter equation into the above equations, the integrated filtration mathematical model of production period is built at last.

6.4.1.2. Physicochemical Parameter Equation
6.4.1.2.1. Reduced Alkaline Concentration
The alkalis used in combination flooding include NaOH, Na_2CO_3, Na_2SiO_3, Na_4SiO_4 and Na_3PO_4, etc.; and OH^- after dissociation plays an important role in water solution.

For NaOH, OH^- concentration in water solution could be given directly.

For Na_2CO_3, OH^- concentration could be deduced by the chemical equilibria shown in Equations 6.4.5 and 6.4.6

$$Na_2CO_3 \xrightarrow{K_1} 2Na^+ + CO_3^{2-} \qquad (6.4.5)$$
$$CO_3^{2-} + 2H_2O \xrightarrow{K_2} H_2CO_3 + 2OH^- \qquad (6.4.6)$$

For the other alkalis, OH^- concentration could be deduced in the same way, and equilibrium constant K_1, K_2 could be found in a chemistry handbook.

6.4.1.2.2. Alkaline Consumption

Considering the fact that quick alkaline consumption resulted from solution and ion exchange on rock surface (similar description as Langmuir adsorption isotherm ion exchange equation) and slow alkaline consumption resulted from rock dissolution, the equation is expressed as follows:

$$R_{OH} = (1 - \phi)\rho_R \frac{\partial c_{rOH}}{\partial t} + \phi \overline{K} c_{OH} \qquad (6.4.7)$$

and

$$c_{rOH} = c_{rOH}^0 + \frac{\alpha c_{OH}}{1 + \beta c_{OH}} \qquad (6.4.8)$$

where

R_{OH} is the alkaline consumption with different alkaline concentrations (g / g);
c_{rOH} is the alkaline consumption mass ratio for ion exchange (g / g);
$\overline{K}$ is the alkaline consumption coefficient (fraction);
c_{rOH}^0 is the minimum alkaline consumption for ion exchange (g / g);
α and β are the constants determined by experiments.

6.4.1.2.3. Surface Active Ion Concentration

The main mechanism of alkaline flooding to enhance oil recovery is to reduce interfacial tension. It can be described as follows: under certain conditions, surfactant is produced when alkaline reacting with acid materials in oil, and then oil and water interfacial tension is reduced. Hence, surface active ion concentration is an important parameter in describing interfacial tension.

$$c_A = \frac{c_{OH}^*}{1 + K / c_{HAo}} \qquad (6.4.9)$$

where:
c_A is the volumetric concentration of surface active ion (mol / l);
c_{OH}^* is the volumetric concentration of alkaline solution injected (mol / l);
c_{HAo} is the concentration of oil acid component, which is a constant corresponding to one kind of volumetric concentration of alkaline injected (mol / l);
K is the reaction coefficient between oil and alkaline, which is a constant corresponding to a certain pH, (mol / l).

6.4.1.2.4. Residual Oil Saturation

Residual oil saturation decreases to some extent with the decreasing interfacial tension. The change in residual oil saturation depends on surface active ion concentration, namely:

$$S_{or}(c_A) = a_1 + a_2 \cdot \exp\left(\frac{-c_A}{a_3}\right) \qquad (6.4.10)$$

where:

$S_{or}(c_A)$ is the residual oil saturation, fraction;

a_1, a_2, a_3 are constants determined by experiments.

6.4.1.2.5. Relative Permeability

The relative permeability curve varies with the change in residual oil saturation; and relative permeability of each phase is expressed as follows:

$$K_{rw} = K_{rw}^o \cdot S^{e_w} \qquad (6.4.11)$$

$$K_{ro} = K_{ro}^o \cdot (1 - S)^{e_o} \qquad (6.4.12)$$

$$S = \frac{S_w - S_{wc}}{1 - S_{or}(c_A) - S_{wc}} \qquad (6.4.13)$$

where:

K_{rw}^o and K_{ro}^o are end-point values on relative permeability curves of water and oil phases, respectively (dimensionless);

e_w and e_o are the relative permeability curve indexes of water and oil phases, respectively, which are modified through linear interpolation with different residual oil saturations;

S is the normalized saturation (fraction);

S_{wc} is the irreducible water saturation (fraction).

6.4.1.2.6. Polymer Solution Viscosity

Polymer solution is a non-Newtonian fluid, whose viscosity is the function of shear rate, and viscosity is influenced by polymer concentration, salinity and shearing. The relationship between polymer solution viscosity and shear rate is given by the Meter and Bird formula:

$$\mu_{wp} = \mu_w + \frac{\mu_0 - \mu_\infty}{1 + \left(\gamma/\gamma_{1/2}\right)^{P_\alpha - 1}} \qquad (6.4.14)$$

where:

μ_w is the water viscosity without polymer (mPa•s);

μ_0 is the zero shear viscosity of polymer solution (mPa•s);

μ_∞ is the infinite shear viscosity, approximated to water viscosity (mPa•s);

$\gamma_{1/2}$ is the shear rate under the polymer viscosity that is in the middle of zero shear viscosity and infinite shear viscosity, it is a constant for certain concentration (s^{-1});

P_α is the correlation factor of polymer solution viscosity and shear rate, it is a constant for certain polymer concentration (dimensionless).

6.4.1.2.7. Determination of Shear Rate

When polymer solution flows through porous media, the shear rate is dependent on polymer solution type, concentration and porous medium. Here, the shear rate formula is built on the derivation of flat plate model:

$$\gamma = \left(\frac{1+3n}{4n}\right)\frac{|\bar{u}|}{\sqrt{8KK_{rw}/\phi S_w}} \qquad (6.4.15)$$

where:

γ is the shear rate (s^{-1});
$\bar{u}$ is the Darcy velocity vector;
n is the Power law exponent.

The Power law exponent is used to describe the derivation extent to Newtonian fluid and varies from 0 to 1.

6.4.1.2.8. Polymer Adsorption

For polymer adsorption, we consider that chemical materials flow through porous media and adsorb on rock surfaces, and the adsorption is irreversible and balance is established instantaneously. Polymer adsorption can be modeled by the Langmuir adsorption isotherm:

$$a\,\widehat{c}_p = \frac{ac_p}{1+bc_p} \qquad (6.4.16)$$

where a and b are constants determined by experiments.

6.4.1.2.9. Permeability Decreasing Factor

When polymer flows through porous media, the permeability decreasing factor is defined as the ratio of water effective permeability to polymer solution effective permeability, which is described as follows:

$$R_K = 1 + \frac{(R_{Kmax}-1)\cdot d\cdot c_p}{1+d\cdot c_p} \qquad (6.4.17)$$

where:

R_{Kmax} is the maximum permeability decreasing factor tested in an experiment, which is constant for a certain polymer;
d is the permeability decreasing factor, which is constant for a certain polymer.

6.4.1.2.10. Resistance Factor and Residual Resistance Factor

Resistance factor is defined as the mobility ratio when water and polymer solution flow through porous media, respectively, which is expressed in the following:

$$R_F = \frac{\lambda_w}{\lambda_p} = \frac{K_w}{K_p}\frac{\mu_p}{\mu_w} = R_K\frac{\mu_p}{\mu_w} \qquad (6.4.18)$$

where:

R_F is the resistance factor;
K_w and K_p are the water and polymer permeabilities, respectively (μm^2).

Residual resistance factor is defined as water mobility ratio before and after polymer solution flow through porous media, which is expressed as follows:

$$R_{RF} = \frac{\lambda_{wi}}{\lambda_{wa}} \tag{6.4.19}$$

where:

R_{RF} is the residual resistance factor;
λ_{wi} and λ_{wa} are the water mobility ratios before and after polymer solution flow through porous media.

6.4.1.2.11. Polymer Inaccessible Pore Volume Fraction

Polymer inaccessible pore volume fraction is approximately described as follows:

$$IPV = (\phi - \phi_p)/\phi \tag{6.4.20}$$

where:

IPV is the inaccessible pore volume fraction (fraction);
ϕ_p is the polymer accessible porosity (fraction).

For the filtration mathematical model of the production period, the streamline method is used to solve saturation, polymer concentration and alkaline concentration, and the resolution of saturation and polymer concentration is the same with the above. In order to solve alkaline concentration, we should convert the alkaline component mass-transfer equation based on grids (Equation 6.4.3) to a one-dimensional equation along the streamline (Equation 6.4.21), and then use the difference method. The solving method is similar to the polymer concentration equation; therefore, there is no more detailed discussion here.

$$\frac{\partial \left[K_a \cdot \frac{(1-\phi)}{\phi} \cdot S_w c_{OH} \right]}{(1 + K_b \cdot c_{OH})^2 \partial t} + \frac{\partial(S_w c_{OH})}{\partial t} + \frac{\partial(c_{OH} f_w)}{\partial \tau} = 0 \tag{6.4.21}$$

6.4.2. Streamline Mathematical Model and Resolution of the Testing Period

6.4.2.1. Pressure Control Equation

Assume total number of streamlines from testing well is N and the flow control equation along each streamline is described as below:

$$\frac{1}{l_j} \frac{\partial}{\partial l_j} \left(l_j \frac{\partial p_j}{\partial l_j} \right) = \frac{\phi C_t}{\lambda_t} \frac{\partial p_j}{\partial t} (j = 1, 2, L, N) \tag{6.4.22}$$

where:

l_j is the curvilinear coordinate along streamline j, the initial point of the curvilinear coordinate system is defined as the position of testing well (cm);
p_j is the pressure on streamline j (10^{-1} MPa);

C_t is the total compressibility factor (10^{-1}MPa);
λ_t is the oil and water phase total mobility [$\mu m^2 / (mPa \bullet s)$].

6.4.2.2. Inner Boundary Condition

In the model, inner boundary condition is divided into production well and injection well inner boundary conditions, respectively; the equations are shown as follows:

$$\left(l_j \frac{\partial p_j}{\partial l_j} \right)_{l_j=r_w} = -\frac{1}{2\pi h \lambda_t} \left\{ q_l - C \left[\frac{dp_w}{dt} - Sl_j \frac{d}{dt} \left(\frac{\partial p_j}{\partial l_j} \right) \right] \right\} (j = 1, 2, L, N)$$

(6.4.23)

$$\left(l_j \frac{\partial p_j}{\partial l_j} \right)_{l_j=r'_w} = -\frac{1}{2\pi h \lambda_t} \left\{ q_w + C \left[\frac{dp_w}{dt} - Sl_j \frac{d}{dt} \left(\frac{\partial p_j}{\partial l_j} \right) \right] \right\} (j = 1, 2, L, N)$$

(6.4.24)

where:

p_w is the pressure of the testing well bore, i.e. the pressure of the first node along each streamline (10^{-1} MPa);
C is the well bore storage coefficient [$cm^3 / (10^{-1}MPa)$];
S is the skin factor (dimensionless);
r_w is the well bore radius of testing well (cm);
r'_w is the oil (water) well bore radius of streamline end-point (cm).

6.4.2.3. Outer Boundary Condition

This is the same as the streamline mathematical model of the testing period for a waterflooding multi-layer reservoir.

6.4.2.4. Initial Condition

This is obtained from the resolution of the filtration mathematical model of the production period for an alkaline / polymer combination flooding multi-layer reservoir.

6.4.2.5. Saturation Equation

$$\frac{\partial S_w}{\partial t} + \frac{\partial f_w}{\partial \tau} = 0$$

(6.4.25)

where:

τ is the propagation time (s);
f_w is the water cut (fraction).

6.4.2.6. Polymer Concentration Equation

$$\frac{\partial\left[F_p\rho_R(\frac{1-\phi}{\phi})S_w\overline{c}_p\right]}{\partial t} + \frac{\partial(S_w c_p)}{\partial t} + \frac{\partial(c_p f_w)}{\partial \tau} = 0 \qquad (6.4.26)$$

6.4.2.7. Alkaline Concentration Convection Mass Transfer Diffusion Equation

$$\frac{\partial\left[K_a \cdot \frac{(1-\phi)}{\phi} \cdot S_w c_{OH}\right]}{(1 + K_b \cdot c_{OH})^2 \partial t} + \frac{\partial(S_w c_{OH})}{\partial t} + \frac{\partial(c_{OH} f_w)}{\partial \tau} = 0 \qquad (6.4.27)$$

The solving method of the streamline mathematical model in the testing period is the same as that for a polymer flooding reservoir, and is solved together with all streamlines on all layers of the testing well. Meanwhile, the calculation of the mobility parameter should consider not only variation of saturation and polymer concentration, but also variation in alkaline concentration (realized by solving Equation 6.4.27).

6.5. COMPARATIVE ANALYSIS OF WELL TESTING PRESSURE RESPONSE CHARACTERISTICS WITH DIFFERENT FLOODING METHODS

As above, influencing characteristics of polymer injection concentration to pressure response in the streamline numerical well testing interpretation model for a polymer flooding reservoir has been studied. In this part, an inverted five-spot typical model will be built in a single-layer reservoir; next, the influence of alkaline injection concentration on model pressure response will be studied; and then the pressure response characteristics of different flooding methods (alkaline flooding, polymer flooding, alkaline / polymer combination flooding) will be compared.

Basic parameters of reservoir and wells: reservoir length is 2000 m; reservoir width is 1500 m; reservoir thickness is 2 m; porosity is 0.2; permeability is 0.8 μm²; initial pressure is 13.5 MPa; initial oil saturation is 0.35; oil and water viscosities are 45 mPa•s and 0.45 mPa•s, respectively; oil and water densities are 0.9 g / cm³ and 1 g / cm³, respectively; total mass production rate of the production well is 60 m³ / d; injection rate of the injection well is 300 m³ / d.

Main physical parameters of alkali and polymer: $K_a = 0.3$; $K_b = 50\,g/g$; $F_P = 0.75$; $\rho_R = 2.63\,g/cm^3$; $c_{rOH}^0 = 130 \times 10^{-6}\,g/g$; $\alpha = 1$, $\beta = 400\,g/g$; $K = 5$; $a_1 = 0.05$, $a_2 = 0.13$, $a_3 = 1.18 \times 10^{-3}\,mol/l$; $P_\alpha = 1$; $R_{Kmax} = 3$; $IPV = 0.25$.

Concentration of injected alkali is 0 (the model is degenerated to water flooding), 1000 mg / l, 3000 mg / l, 6000 mg / l, 9000 mg / l, 12,000 mg / l and 15,000 mg / l, respectively, shut-in injection well after alkaline flooding for 300 days, and then take the pressure drawdown test. The pressure and its derivative log–log curve are shown in Fig. 6.17.

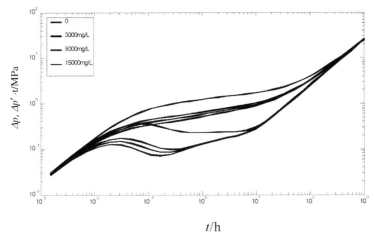

FIGURE 6.17 Pressure response contrast with different injection alkaline concentrations.

It can be seen in Fig. 6.17 that, under different injection alkaline concentrations conditions, the higher the alkaline concentration is, the lower the height of the pressure derivative curve middle section will be. This is because oil/water interfacial tension is reduced during the alkaline flooding process, and residual oil saturation has been changed. Within certain limits, the higher the injection alkaline concentration is, the greater the total mobility will be.

With a constant working system, shut-in the injection well after alkaline flooding (injection concentration 6000 mg/l), polymer flooding (injection concentration 1500 mg/l) and combination flooding (alkaline injection concentration 6000 mg/l, polymer injection concentration 1500 mg/l) for 300 days, respectively, and then pressure drawdown simulation is calculated to get the pressure and its derivative log–log curve with different displacement modes, as shown in Fig. 6.18.

As shown in Fig. 6.18, in the case of alkaline flooding, it has low oil saturation and highest total mobility around the injection well. Hence, the height of the pressure derivative curve is the lowest. While keeping away from the testing well, total mobility decreases and the pressure derivative curve rises gradually. In the case of polymer flooding, the pressure derivative curve drops after the radial flow period; this is because total mobility increases when pressure passes through the polymer region and comes into the oil region.

From the contrast with the pressure response curve of alkaline flooding, polymer flooding and alkaline/polymer combination flooding, it could be concluded that: there is not much difference between alkaline/polymer combination flooding and polymer flooding for the pressure response curve, which both show the pressure response characteristics of a polymer flooding reservoir. This is because the influence of polymer on fluid mobility is much greater than that of

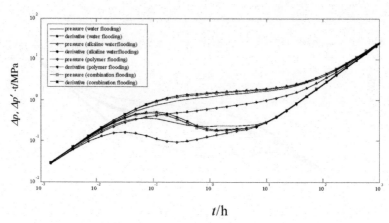

FIGURE 6.18   Pressure response contrast of water alkaline, polymer and combination flooding.

alkali, and the influence characteristics of alkali on pressure response is negligible compared with polymer.

6.6. CHAPTER SUMMARY

In this chapter, based on the streamline numerical well testing interpretation model for a waterflooding multi-layer sandstone reservoir, the streamline numerical well testing interpretation models for polymer flooding and alkaline/polymer combination flooding multi-layer sandstone reservoirs are studied, and the main conclusions are shown as below.

1. Streamline numerical well testing interpretation models for polymer flooding and alkaline/polymer combination flooding multi-layer sandstone reservoir include the filtration mathematical model of the production period and the streamline mathematical model of the testing period.
   The filtration mathematical model of the production period uses a simplified polymer flooding and alkaline/polymer combination flooding multi-layer sandstone reservoir black oil model, which could consider the main factors of polymer flooding (viscosity variation, adsorption, permeability decrease, and inaccessible pore volume, etc.) and alkaline flooding (alkaline concentration, alkaline consumption, surface active ion concentration, residual oil saturation); the mathematical model of the testing period is the streamline model, which is built by combining together the mathematical equations for each streamline on each layer of the testing well. Pressure, saturation and polymer concentration, which are obtained from the filtration mathematical model, are used as the basic and initial conditions for the streamline mathematical model of the testing period.
2. A combination of the IMPES and the streamline methods is used to solve the filtration mathematical model of the production period. The IMPES method

is used to solve pressure distribution, and the streamline method is used to solve saturation and polymer concentration distribution, and the main procedures are shown as follows:

- the IMPES method is used to solve the pressure control equation and physicochemical parameter equation, and then the pressure distribution of the studying region will be obtained;
- on the basis of pressure distribution, trace the streamline and obtain the streamline distribution of the studying region;
- convert the grid parameter to node parameter on the streamline, and build the saturation and polymer concentration equation along the streamline;
- solve the saturation and polymer concentration equation along the streamline, and obtain the saturation and polymer concentration distribution at the new time;
- if there is great variation in streamline distribution and node parameter, use the IMPES method to solve pressure and then update the streamline;
- build and solve the saturation and polymer concentration equation along the new streamline, and obtain the saturation and polymer concentration distribution of the testing period.

Resolution of the streamline mathematical model of the testing period is similar to that of water flooding. However, there are a few differences, as follows: node mobility in the streamline mathematical model of the testing period for a polymer flooding multi-layer sandstone reservoir is related not only to absolute permeability and saturation, but also to polymer concentration, so the polymer concentration equation and polymer physicochemical parameter equation should be combined together to solve and update mobility parameter on new streamline nodes continuously.

3. With pressure drawdown simulation of the testing well in a typical model for a polymer flooding multi-layer sandstone reservoir, the pressure response characteristics of the testing well during the process of polymer flooding in different periods are studied:

- While testing the wells by injecting polymer, streamline distribution shows two regions with testing well as the center, namely the polymer slug region and the oil region. When the pressure response of the testing well enters the oil region from the polymer slug region, the pressure derivative curve sinks due to the mobility increase.
- While testing the well with water flooding after polymer injection, streamline distribution shows three regions with testing well as the center, namely the water region, the polymer slug region and the oil region. The mobility of the water region is the biggest and the mobility of the polymer slug region is the smallest; when the pressure response of the testing well enters the polymer slug region from the water region, mobility decreases and the pressure derivative curve rises for the first time; when the pressure response enters the oil region from the polymer slug region, the mobility increases and the pressure derivative curve sinks;

because of layer permeability differences, the pressure response of the testing well enters the polymer slug region from the water region at a different time, and the slope of the first pressure derivative rising varies continuously.

4. Pressure response influencing characteristics of the streamline numerical well testing interpretation model with different displacement modes are compared and analyzed.

- In the situation of alkaline flooding, with different injection alkaline concentrations, the greater the injection alkaline concentration is, the lower the height of the pressure derivative curve in the middle part will be;
- In the situation of alkaline/polymer combination flooding, there is not much difference between the pressure response curve of alkaline/polymer combination flooding and polymer flooding, and they both show the pressure response characteristics of a polymer flooding reservoir. This is because the influence of polymer on fluid mobility is much greater than that of alkali, and the influence characteristics of alkaline on pressure response is concealed by polymer.

Streamline Numerical Well Testing Interpretation Model Considering Components

For some special reservoirs, such as volatile reservoirs, condensate gas reservoirs, and gas injection miscible development reservoirs, phase variation of sub-surface fluid always occurs during the production process. For these kinds of reservoirs, sub-surface fluid flow is a complex, strongly non-linear flow problem, and there are many influencing factors, such as composition of sub-surface fluid, composition of injection fluid, injection time, location and range of phase variation; the curve-matching method used for vertical wells can not be used for well testing interpretation of this kind of reservoir. Hence, a numerical well testing interpretation simulator taking into consideration underground components will be built in this chapter, which can simulate testing well pressure response with phase variation of sub-surface fluids.

7.1. COMPOSITIONAL MODEL

7.1.1. Basic Assumption

During the development process of a special reservoir, phase variation and miscible phase often happen among sub-surface fluids, and these fluid variations have great influence on pressure testing. The compositional model can consider phase variations, and assumptions for the model are as follows:

1. The reservoir is homogeneous and anisotropic.
2. Before reservoir development, the formation pressure distributes uniformly with p_i; reservoir temperature is T and is constant during the development process. Sub-surface fluid is single phase (oil phase) in the initial conditions.
3. At any time during the development process, thermodynamic equilibrium is finished instantaneously when the phase variation of oil/gas phases occurs in the reservoir.
4. The reservoir pressure gradient is small during the production process.
5. The influence of capillary pressure and gravity are considered.
6. Sub-surface fluid flow follows the Darcy law.

Streamline Numerical Well Test Interpretation. DOI: 10.1016/B978-0-12-386027-9.00007-1
Copyright © 2011 by Elsevier Ltd

7.1.2. The Compositional Model

The compositional model includes the filtration model porous flow equation of the water phase and the oil/gas component, the phase equilibrium equation, the auxiliary equation, inner/outer boundary conditions, and so on.

7.1.2.1. Filtration Equation of the Water Phase

The kinematic equation is obtained from the mass conservation equation, the Darcy equation and the state equation:

$$\nabla \cdot [\lambda_w \rho_w \nabla \Phi_w] + q_w = \frac{\partial}{\partial t}(\phi \rho_w S_w) \qquad (7.1.1)$$

7.1.2.2. Filtration Equation of Hydrocarbon Component i

The filtration equation of component i is obtained from the mass conservation equation, the Darcy equation and the state equation of component i:

$$\nabla \cdot [\lambda_o \rho_o x_i \nabla \Phi_o + \lambda_g \rho_g y_i \nabla \Phi_g] + q_i = \frac{\partial}{\partial t}[\phi F z_i] \qquad (7.1.2)$$

7.1.2.3. Total Hydrocarbon Equation

$$\nabla \cdot [\lambda_o \rho_o \nabla \Phi_o + \lambda_g \rho_g \nabla \Phi_g] + q_h = \frac{\partial}{\partial t}[\phi F] \qquad (7.1.3)$$

Where

z_i = total mole fraction of component i in hydrocarbon system;
x_i = mole fraction of component i in liquid phase;
y_i = mole fraction of component i in gas phase;

$$F = \rho_o S_o + \rho_w S_w;$$

ϕ = porosity;
ρ_o, ρ_g and ρ_w = mole density of oil, gas and water phase respectively;
o,g,w = oil, gas and water phase respectively.

$$\Phi_o = p_o - \gamma_o D$$
$$\Phi_o = p_g - \gamma_g D = p_o + p_{cgo} - \gamma_o D$$
$$\Phi_w = p_w - \gamma_w D = p_o - p_{cwo} - \gamma_w D$$

Here, p_o, p_g and p_w are the pressure of oil, gas and water phases, respectively; γ_o, γ_g and γ_w are the gravity of oil, gas and water phases, respectively; D is the depth; λ_o, λ_g and λ_w are the flow coefficients of oil, gas and water phases, respectively. The definition formula is shown as follows:

$$\lambda_l = \frac{K K_{rl}}{\mu_l} \qquad (l = o, g, w)$$

ϕ, ρ_o, ρ_g and ρ_w are determined in the following formula:

$$\phi = \phi^* \cdot \left[1 + C_f\left(p_o - p_f\right)\right]$$

$$\rho_w = \rho_w^* \cdot \left[1 + C_w(p_o - p_w)\right]$$

$$\rho_g = \frac{p_g}{z_g RT}$$

$$\rho_o = \frac{p_o}{z_o RT}$$

Where ϕ^* is porosity at pressure p_{wf}; C_f is rock compressibility factor; ρ_w^* is water density at initial base pressure p_w; R is universal gas constant (82.06 atm.ml / mol.k); z_g and z_o is compressibility factor of gas and oil respectively.

7.1.2.4. Equilibrium Equation

Thermodynamic equilibrium condition: when the oil/gas system achieves equilibrium, fugacity of any component i in oil and gas phase should be equal, that is:

$$f_i^L = f_i^V \qquad (i = 1, 2, \cdots, N_c) \tag{7.1.4}$$

For real gas:

$$\psi_i^L x_i = \psi_i^V y_i \qquad (i = 1, 2, \cdots, N_c) \tag{7.1.5}$$

where f is component fugacity; ψ is fugacity factor of real gas; L is liquid phase and V is gas phase. N_c is component number.

The following algebraic equation is obtained from the material balance equation of the oil/gas phase:

$$z_i = Lx_i + Vy_i \qquad (i = 1, 2, \cdots, N_c) \tag{7.1.6}$$

where L is mole fraction of liquid phase in hydrocarbon system; V is mole fraction of gas phase in hydrocarbon system.

$$L = \left(\frac{\rho_o S_o}{\rho_o S_o + \rho_g S_g}\right) = \left(\frac{\rho_o S_o}{F}\right) \tag{7.1.7}$$

$$V = \left(\frac{\rho_g S_g}{\rho_o S_o + \rho_g S_g}\right) = \left(\frac{\rho_g S_g}{F}\right) \tag{7.1.8}$$

At pressure p_o, fugarity f_i^L, f_i^V and mole density ρ_o, ρ_g can be calculated.

7.1.2.5. Auxiliary Equations

The state equation, capillary pressure equation and saturation equation are mainly included as follows:

$$\sum_{i=1}^{N_c} x_i = \sum_{i=1}^{N_c} y_i = \sum_{i=1}^{N_c} z_i = 1 \tag{7.1.9}$$

$$L + V = 1 \tag{7.1.10}$$

$$p_o - p_w = p_{cwo}(s_w) \tag{7.1.11}$$

$$p_g - p_o = p_{cog}(s_g) \tag{7.1.12}$$

$$s_w + s_o + s_g = 1 \tag{7.1.13}$$

$$S_o + F\left(\frac{L}{\rho_o} + \frac{1-L}{\rho_g}\right) = 1 \tag{7.1.14}$$

7.1.2.6. Definite Condition

For specific questions, initial conditions and boundary conditions need to be provided.

7.1.2.6.1. Initial Conditions

Initial conditions include the original distribution of pressure, saturation and composition in the oil/gas reservoir at the initial time or from a certain time of exploitation.

$$p_w(r,t)\big|_{t=0} = p_{wi} \qquad p_o(r,t)\big|_{t=0} = p_{oi} \qquad p_g(r,t)\big|_{t=0} = p_{gi} \tag{7.1.15}$$

$$s_w(r,t)\big|_{t=0} = s_{wi} \qquad s_o(r,t)\big|_{t=0} = s_{oi} \qquad s_g(r,t)\big|_{t=0} = s_{gi} \tag{7.1.16}$$

$$z_i(r,t)\big|_{t=0} = z_{ii} \tag{7.1.17}$$

7.1.2.6.2. Outer Boundary Conditions

Boundary conditions includes two types: closed boundary and constant pressure boundary.

For closed boundary:

$$\frac{\partial \Phi_o}{\partial n}\bigg|_r = 0 \quad \frac{\partial \Phi_g}{\partial n}\bigg|_r = 0 \quad \frac{\partial \Phi_w}{\partial n}\bigg|_r = 0 \tag{7.1.18}$$

For constant pressure boundary:

$$p\big|_r = const \tag{7.1.19}$$

7.1.2.6.3. Inner Boundary Conditions

For inner boundary conditions, constant rate and constant bottom-hole pressure are considered. For specific conditions, the variation of production well or injection well rate in different periods are also considered. For constant bottom-hole pressure,

$$p\big|_{r=r_w} = const \tag{7.1.20}$$

For constant well flow rate,

$$q\big|_{r=r_w} = const \tag{7.1.21}$$

7.2. STATE EQUATION AND PHASE EQUILIBRIUM

Phase state variation happens when fluid flows underground, and phase state is described through phase equilibrium. Although oil/gas hydrocarbon stays at thermal equilibrium state in the oil/gas reservoir while building the compositional model, the components in hydrocarbon system show liquid, gas, or co-existing states; liquid and gas phases have different flow characteristics. Hence, it is very important to study the phase state and variation of components during the flowing process.

7.2.1. The PR State Equation

Since Van der Waal's state equation was published in 1873, almost 100 state equations have been proposed. The PR state equation is used in our work.

The PR state equation was proposed in 1976, which had a positive effect on estimating the phase state of multi-component mixture systems. The PR state equation has two constants and is easy to use; it also estimates the gas density equal volume behavior with good accuracy. Hence, the PR state equation is widely used in the oil and gas industries.

The form of the PR state equation is described as follows:

$$p = \frac{RT}{V-b} - \frac{\alpha(T)}{V(V+b)+b(V-b)} \tag{7.2.1}$$

where R is the gas constant; T is temperature; V is mole volume; b is real volume; constant $\alpha(T)$ is the function of temperature, which is described as follows:

$$\alpha(T) = \alpha(T_c)\, \alpha\,(T_r, \omega)$$
$$\alpha(T_c) = 0.45727 R^2 T_c^2 / p_c$$

$\alpha(T_r, \omega)$ is a dimensionless function of reference temperature T_r and eccentricity factor ω

$$[\alpha(T_r, \omega)]^{0.5} = 1 + K'(1 - T_r)^{0.5}$$

K' is the inherent characteristic constant of every material, which is determined by:

$$K' = 0.37464 + 1.54226\omega - 0.26992\omega^2$$

b in Equation 7.2.1:

$$b = \sum x_i b_i$$

Cubic equation for state equation:

$$Z^3 - (1-B)Z^2 + (A - 2B - 3B^2)Z - (AB - B^2 - B^3) = 0 \qquad (7.2.2)$$

where

$$Z = \frac{pV}{RT}$$

$$A = \frac{ap}{R^2 T^2}$$

$$B = \frac{bp}{RT}$$

Miscible phase rule:

$$a = \sum_i \sum_j x_i x_j a_{ij}$$

$$a_{ij} = (1 - K_{ij})(a_i a_j)^{0.5}$$

$$a_i = \Omega_{ai} \frac{R^2 T_{ci}^2}{p_{ci}}$$

$$\Omega_{ai} = 0.45724$$

$$a_i^{0.5} = 1 + \lambda_i \left[1 - \left(\frac{T}{T_{ci}}\right)^{0.5}\right]$$

$$\lambda_i = 0.375 + 1.542\omega_i - 0.27\omega_i^2$$

$$b = \sum_i x_i b_i$$

$$b_i = \Omega_{bi} \frac{RT_{ci}}{p_{ci}}$$

$$\Omega_{bi} = 0.0778$$

Fugacity equation:

$$\ln\frac{f_i}{px_i} = \frac{b_i}{b}(Z - 1) - \ln(Z - b) - \frac{1}{2\sqrt{2}}\frac{A}{B}\left(\frac{2\sum x_j a_{ij}}{a} - \frac{b_i}{b}\right)\ln\frac{Z + 2.414B}{Z - 0.414B}$$

Critical volume:

$$V_{ci} = 0.30740\frac{RT_{ci}}{p_{ci}}$$

7.2.2. Phase Equilibrium Calculation

The phase equilibrium calculation is based on the principle that gas/liquid fugacity is equivalent in the equilibrium condition; reservoir fluid is assumed to be a multi-component non-ideal system, then gas/liquid equilibrium, bubble point and dew point can be calculated as follows.

7.2.2.1. Total Mass Conservation Equation

Assume the equilibrium of gas and liquid phase, the fluid total mole number in system is 1, and $L + V = 1$ where L is the liquid phase mole number at equilibrium condition; V is the gas phase mole number at equilibrium condition.

7.2.2.2. Compositional Mass Conservation Equation

$$Lx_i + Vy_i = z_i$$

where x_i is the mole fraction of component i in liquid phase; y_i is the mole fraction of component i in gas phase; z_i is the mole fraction of component i in the system.

7.2.2.3. Constraint Equation of Phase Components

$$\sum_{i=1}^{N_c} x_i = \sum_{i=1}^{N_c} \left(\frac{z_i}{L + K_i V} \right) = 1 \tag{7.2.3}$$

$$\sum_{i=1}^{N_c} y_i = \sum_{i=1}^{N_c} \left(\frac{z_i}{L/K_i + V} \right) = 1 \tag{7.2.4}$$

where K_i is gas/liquid equilibrium constant of component i, $K_i = y_i/x_i$.

7.2.2.4. Fugacity Equation

For ideal gas:

$$f_i^L = f_i^V \tag{7.2.5}$$

According to thermodynamic principles, while gas and liquid phases stay at phase state equilibrium, the fugacity of each component in each phase is equivalent; the concept of fugacity is introduced for real gas $\psi = \frac{f}{p}$, ψ is fugacity factor and fugacity f can be seen as corrected pressure. Hence, the fugacity equation of real gas is shown in the following:

$$\psi_i^L p_i^L = \psi_i^V p_i^V$$

or

$$\psi_i^L x_i p = \psi_i^V y_i p \tag{7.2.6}$$

7.2.2.5. Flash Calculation

Flash calculation means to calculate the phase composition x_i, y_i, liquid phase mole fraction L and gas phase mole fraction V, where system pressure p, system temperature T and mixture composition z_i are known.

Actually, if $N_c + 1$ unknown numbers x_i, L (or y_i, V, $i = 1,2,\ldots N_c$) are calculated, $N_c + 1$ unknown numbers of the other phase can be obtained from the mass equilibrium equation. The iterative process to solve the liquid phase unknown numbers is called L–V iteration, which is applicable to the gas phase dominant situation, namely $V > 0.5$; at this situation L is lower and its phase variation is sensitive to equilibrium calculation, taking x_i and L as the resolution variables could improve calculation accuracy and solution convergence. On the contrary, when the liquid phase is dominant, the resolution of y_i, V is called V–Y iteration. The interchangeable methods of solving variables could be implemented automatically by programs.

7.2.2.5.1. Newton–Raphson Flash Calculation Method

The fugacity equation and the phase component constraint equation can be written in the form of error equation:

$$R_i = \psi_i^V y_i - \psi_i^L x_i \qquad (i = 1, 2, \cdots, N_c) \tag{7.2.7}$$

$$R_{N_c+1} = 1 - \sum_{i=1}^{N_c} x_i = 0 \tag{7.2.8}$$

The resolution variables in the equations are x_i, L, and there are $N_c + 1$ equations in Equations 7.2.7 and 7.2.8, so the $N_c + 1$ variables can be solved.

Difference on the $N_c + 1$ equations above forms the following algebraic equations:

$$\vec{J}\,\delta\vec{x} = -\vec{R} \tag{7.2.9}$$

where $\vec{R}$ is remainder term, $R_i(i = 1,2,\ldots,N_c)$; $\delta\vec{x}$ is two iterations difference of the resolution variables; $\vec{J}$ is Jacobian matrix.

$$\vec{J} = \begin{bmatrix} \dfrac{\partial R_1}{\partial x_1} & \dfrac{\partial R_1}{\partial x_2} & \cdots & \dfrac{\partial R_1}{\partial x_{N_c}} & \dfrac{\partial R_1}{\partial L} \\[2mm] \dfrac{\partial R_2}{\partial x_1} & \dfrac{\partial R_2}{\partial x_2} & \cdots & \dfrac{\partial R_2}{\partial x_{N_c}} & \dfrac{\partial R_2}{\partial L} \\[2mm] \vdots & \vdots & \vdots & \vdots & \vdots \\[2mm] \dfrac{\partial R_{N_c}}{\partial x_1} & \dfrac{\partial R_{N_c}}{\partial x_2} & \cdots & \dfrac{\partial R_{N_c}}{\partial x_{N_c}} & \dfrac{\partial R_{N_c}}{\partial L} \\[2mm] \dfrac{\partial R_{N_c+1}}{\partial x_1} & \dfrac{\partial R_{N_c+1}}{\partial x_2} & \cdots & \dfrac{\partial R_{N_c+1}}{\partial x_{N_c}} & \dfrac{\partial R_{N_c+1}}{\partial L} \end{bmatrix} \quad \delta\vec{x} = \begin{bmatrix} \delta x_1 \\ \delta x_2 \\ \vdots \\ \delta x_{N_c} \\ \delta x_{N_c+1} \end{bmatrix} \quad \vec{R} = \begin{bmatrix} R_1 \\ R_1 \\ \vdots \\ R_{N_c} \\ R_{N_c+1} \end{bmatrix}$$

Because y_i is free variable, derivation of y_i can be converted to derivation o x_i and L. According to conservation equilibrium:

$$z_i = Lx_i + (1 - L)x_i \tag{7.2.10}$$

$$y_i = \frac{z_i - Lx_i}{1 - L} \tag{7.2.11}$$

$$\frac{\partial y_i}{\partial x_i} = \frac{L}{L-1} \tag{7.2.12}$$

Hence we have the following formula:

$$\frac{\partial R_i}{\partial x_j} = \frac{L}{L-1}\frac{\partial}{\partial y_j}(y_i\psi_i^V) - \frac{\partial}{x_j}(x_i\psi_i^L) = \frac{L}{L-1}\left(\frac{\partial y_i}{\partial y_j}\psi_i^V + y_i\frac{\partial \psi_i^V}{\partial y_j}\right)$$
$$- \left(\frac{\partial x_i}{\partial x_j}\psi_i^L + x_i\frac{\partial \psi_i^L}{\partial x_j}\right) \tag{7.2.13}$$

$$\frac{\partial R_i}{\partial L} = \frac{L}{L-1}\sum(x_i - y_i)\left(\frac{\partial y_i}{\partial y_j}\psi_i^V + y_i\frac{\partial \psi_i^V}{\partial y_j}\right) \tag{7.2.14}$$

$$\frac{\partial R_{N_c+1}}{\partial x_j} = -1 \tag{7.2.15}$$

$$\frac{\partial R_{N_c+1}}{\partial L} = 0 \tag{7.2.16}$$

From Formula 7.2.9:

$$\delta \vec{x} = -\vec{J}^{-1}\vec{R} \tag{7.2.17}$$

Newton iteration is used to solve non-linear equations. At first, to eliminate on row vector and convert matrix $\vec{J}$ to triangular matrix line by line, then take it into the equation to get the increment of x_i, L, that is δx_i, δL; based on the new iteration value (with the increment of last step), repeat this process until convergence.

After getting x_i and L, y_i and z_i are solved from Equations 7.2.10 and 7.2.11.

7.2.2.5.2. Successive Displacement Flash Method

(1) Empirical formula is used to study equilibrium ratio K_i:

$$K_i = \frac{p_{ci}}{p}\exp\left[5.3727(1 + \omega_i)\left(1 - \frac{T_{ci}}{T}\right)\right] \tag{7.2.18}$$

(2) With K_i and z_i known, flash method is used to solve L

$$z_i = Lx_i + (1-L)y_i = [L + (1-L)K_i]x_i \tag{7.2.19}$$

$$\sum_{i=1}^{N_c}(x_i - y_i) = \sum_{i=1}^{N_c}x_i(1 - K_i) = 0 \tag{7.2.20}$$

$$\sum_{i=1}^{N_c}\frac{z_i(1 - K_i)}{L + (1-L)K_i} = 0 \tag{7.2.21}$$

L could be obtained from Formula 7.2.21.

(3) With z_i, K_i and L known, x_i and y_i are solved:

$$x_i = \frac{z_i}{L + (1 - L)K_i} \qquad (7.2.22)$$
$$y_i = K_i \cdot x_i$$

(4) Based on x_i and y_i, K_i is recalculated:

$$K_i = \frac{\psi_i^L}{\psi_i^V} \qquad (7.2.23)$$

(5) Accelerated GDEM method:

$$\ln K_{i,j+1} = \ln K_{i,j} + \frac{\ln K_{i,j} - \ln K_{i,j-1}}{1 + M} \qquad (7.2.24)$$

where

$$M = \left(\ln \frac{K_{i,j}}{K_{i,j-1}}, \ln \frac{K_{i,j-1}}{K_{i,j-2}} \right) \Big/ \left(\ln \frac{K_{i,j-1}}{K_{i,j-2}}, \ln \frac{K_{i,j-1}}{K_{i,j-2}} \right) \qquad (7.2.25)$$

Notes: around the critical point $M < 0$, the RISNES method is used to accelerate.

RISNES method:

$$K_{i,j+1} = K_{i,j} R_{i,j}^{1/(1-k)} \qquad (7.2.26)$$

where

$$k = \frac{R_j - 1}{R_{j-1} - 1}$$

and

$$R_{i,j} = \left(\frac{K_i x_i}{y_i} \right)_j$$

Iterate until M, k equals to constant and then accelerate. Choose the method which can get the constant faster. Get K_i and recycle to (2) repeat the process of (2)–(5) until convergence.

7.2.2.5.3. Comparison of the Two Methods

The advantage of the Newton–Raphson method is its fast convergence, but it depends heavily on initial sensitivity. When x_i and L are not given a good initial value, convergence may not be achieved, and the closer to critical point the value is, the stricter the initial value of x_i and L should be given.

The advantages of the successive displacement method are easy calculation and reliability. Convergence happens even around critical point. The disadvantage is slow convergence rate and thousands of iterations maybe happen around critical point.

In this model, the two methods are combined: use the Newton–Raphson method at first and then use the successive iteration method when misconvergence happens. The user can also choose which method to use before calculation.

7.2.2.6. Dew Point and Bubble Point Calculation

The Newton–Raphson method is used for both the dew point and bubble point calculations, and the calculation formula is the same as the Newton–Raphson method in flash calculation. Here, only the calculation formula of the Jacobian matrix element is given.

7.2.2.6.1. Bubble Point Calculation

When phase variation of oil/gas system stays at bubble point state, equilibrium is achieved and mole fraction of the gas phase $V \to 0$, mole fraction of the liquid phase $L \to 1$, as indicated in the following formula:

$$y_i = \frac{z_i K_i}{1 + (K_i - 1)V} = z_i K_i \tag{7.2.27}$$

$$x_i = \frac{z_i}{1 + (K_i - 1)V} = z_i \tag{7.2.28}$$

Accordingly, normalization condition of bubble point components is shown in the following:

$$\sum_{i=1}^{N_c} y_i = \sum_{i=1}^{N_c} z_i K_i = 1 \tag{7.2.29}$$

Hence, Formula 7.2.7 converts as follows:

$$\psi_i^L z_i = \psi_i^V y_i \tag{7.2.30}$$

Expression of Equations 7.2.30 and 7.2.31 in the form of an error equation:

$$R_i = \psi_i^V y_i - \psi_i^L z_i \quad (i = 1, 2 \ldots N_c) \tag{7.2.31}$$

$$R_{N_c+1} = 1 - \sum_{i=1}^{N_c} y_i = 0 \text{ (Iteration for the first ten times)} \tag{7.2.32}$$

$$R_{N_c+1} = 1 - \sum_{i=1}^{N_c} \frac{\psi_i^L}{\psi_i^V} z_i \text{ (Iteration after first ten times)} \tag{7.2.33}$$

The solution variables in equations are y_i and p (bubble point), where: $i = 1, 2, \ldots, N_c$. Then:

$$\frac{\partial R_i}{\partial y_i} = \psi_i^V \frac{\partial y_i}{\partial y_j} + y_i \frac{\partial \psi_j^V}{\partial y_j} \tag{7.2.34}$$

or

$$\frac{\partial R_i}{\partial p} = -z_i \frac{\partial \psi_i^L}{\partial p} + y_i \frac{\partial \psi_i^V}{\partial P} \tag{7.2.35}$$

$$\frac{\partial R_{N_c+1}}{\partial y_j} = -1$$

or

$$\frac{\partial R_{N_c+1}}{\partial y_j} = \sum_{i=1}^{N_c+1} z_i \left[\frac{\psi_i^L}{(\psi_i^V)^2} \cdot \frac{\partial \psi_i^V}{\partial p} - \frac{1}{\psi_i^V} \cdot \frac{\partial \psi_i^L}{\partial p} \right] \tag{7.2.36}$$

7.2.2.6.2. Dew Point Pressure Calculation

When the phase variation of the gas system stays at dew point state, the material equilibrium relationship of the phases show that mole fraction of gas phase $V \to 1$, mole fraction of liquid phase $L \to 0$. Hence,

$$y_i = \frac{z_i K_i}{1 + (K_i - 1)V} = z_i \tag{7.2.37}$$

$$x_i = \frac{z_i}{1 + (K_i - 1)V} = \frac{z_i}{K_i} \tag{7.2.38}$$

Normalization condition of dew point components is shown in the following:

$$\sum_{i=1}^{N_c} x_i = \sum_{i=1}^{N_c} \frac{z_i}{K_i} = 1 \tag{7.2.39}$$

Formula 7.2.7 converts as follows:

$$\psi_i^V z_i = \psi_i^L x_i \tag{7.2.40}$$

Expression of Equations 7.2.39 and 7.2.40 in the form of an error equation:

$$R_i = \psi_i^V z_i - \psi_i^L x_i \quad (i = 1, 2, \dots, N_c) \tag{7.2.41}$$

$$R_{N_c+1} = 1 - \sum_{i=1}^{N_c} x_i = 0 \text{ (Iteration for the first ten times)} \tag{7.2.42}$$

$$R_{N_c+1} = 1 - \sum_{i=1}^{N_c} \frac{\psi_i^V}{\psi_i^L} z_i = 0 \text{ (Iteration after first ten times)} \tag{7.2.43}$$

The solution variables are y_i and p (dew point), where: $i = 1, 2, \dots, N_c$. Hence, the equations are described as follows:

$$\frac{\partial R_i}{\partial y_j} = -\left(\psi_i^L \frac{\partial y_i}{\partial y_j} + y_i \frac{\partial \psi_i^V}{\partial y_j} \right) \tag{7.2.44}$$

or

$$\frac{\partial R_i}{\partial p} = -z_i \frac{\partial \psi_i^L}{\partial p} + y_i \frac{\partial \psi_i^V}{\partial P} \tag{7.2.45}$$

$$\frac{\partial R_{N_c+1}}{\partial x_j} = 0 \tag{7.2.46}$$

or

$$\frac{\partial R_{N_c+1}}{\partial x_j} = \sum_{i=1}^{N_c+1} \frac{\psi_i^V z_i}{(\psi_i^L)^2} \cdot \frac{\partial \psi_i^L}{\partial x_j} \tag{7.2.47}$$

$$\frac{\partial R_{N_c+1}}{\partial p} = 0 \tag{7.2.48}$$

or

$$\frac{\partial R_{N_c+1}}{\partial p} = \sum_{i=1}^{N_c+1} z_i \left[\frac{\psi_i^V}{(\psi_i^L)^2} \cdot \frac{\partial \psi_i^L}{\partial p} - \frac{1}{\psi_i^L} \cdot \frac{\partial \psi_i^V}{\partial p} \right] \tag{7.2.49}$$

7.3. IMPES SOLUTION OF THE COMPOSITIONAL MODEL

7.3.1. Finite Difference Equation

Referring to the difference in the $2N_c + 2$ equations in the above model, solve conductivity explicitly with the IMPES method and obtain the following equations:
    Equation of water component:

$$\Delta T_w^n \Delta \Phi_w^{n+1} + q_w^n = \frac{V}{\Delta t} \left[(\phi \rho_w S_w)^{n+1} - (\phi \rho_w S_w)^n \right] \tag{7.3.1}$$

Equation of component i ($i = 1, 2, \ldots, N_c - 1$):

$$\Delta T_o^n x_i^n \Delta \Phi_o^{n+1} + \Delta T_g^n y_i^n \Delta \Phi_g^{n+1} + q_i^n = \frac{V}{\Delta t} \left[(\phi F z_i)^{n+1} - (\phi F z_i)^n \right] \tag{7.3.2}$$

Total hydrocarbon equation:

$$\Delta T_o^n \Delta \Phi_o^{n+1} + \Delta T_g^n \Delta \Phi_g^{n+1} + q_h^n = \frac{V}{\Delta t} \left[(\phi F)^{n+1} - (\phi F)^n \right] \tag{7.3.3}$$

Fugacity equation:

$$\psi_i^L x_i = \psi_i^V y_i \qquad (i = 1, 2, \cdots, N_c) \tag{7.3.4}$$

Saturation equation:

$$S_w + F \left(\frac{L}{\rho_o} + \frac{1-L}{\rho_g} \right) = 1 \tag{7.3.5}$$

where

$$T_l = \frac{KK_{rl}}{\mu_l} \qquad (l = o, g, w)$$

Expand the above differential equations, transform to the form of iteration remainder and make orders.

7.3.1.1. Fugacity Equation

$$R_i = \psi_i^V y_i - \psi_i^L x_i = 0 \quad (i = 1, 2, \cdots, N_c) \tag{7.3.6}$$

7.3.1.2. Compositional Flow Equation

$$
\begin{aligned}
R_{N_c+i} &= \Delta \left[\frac{TK_{ro}\rho_o x_i}{\mu_o} \left(\Delta p_o^{n+1} - \gamma_o \Delta D \right) + \frac{TK_{rg}\rho_g y_i}{\mu_g} \left(\Delta p_o^{n+1} + \Delta p_{cgo} - \gamma_g \Delta D \right) \right] \\
&- \left(q_o^n x_i + q_g^n y_i \right) - \frac{V}{\Delta t} \left[(\phi F z_i)^{n+1} - (\phi F z_i)^n \right] = 0 \quad (i = 1, 2, \cdots, N_c - 1)
\end{aligned}
\tag{7.3.7}
$$

7.3.1.3. Total Hydrocarbon Equation

$$
\begin{aligned}
R_{2N_c} &= \Delta \left[\frac{TK_{ro}\rho_o}{\mu_o} \left(\Delta p_o^{n+1} - \gamma_o \Delta D \right) + \frac{TK_{rg}\rho_g}{\mu_g} \left(\Delta p_o^{n+1} + \Delta p_{cgo} - \gamma_g \Delta D \right) \right] \\
&- \left(q_o^n + q_g^n \right) - \frac{V}{\Delta t} \left[(\phi F)^{n+1} - (\phi F)^n \right] = 0
\end{aligned}
\tag{7.3.8}
$$

7.3.1.4. Water Flow Equation

$$
\begin{aligned}
R_{N_c+1} &= \Delta \left[\frac{TK_{rw}\rho_w}{\mu_w} \left(\Delta p_o^{n+1} - \Delta p_{cwo} - \gamma_w \Delta D \right) \right] \\
&- q_w^n - \frac{V}{\Delta t} \left[(\phi S_w \rho_w)^{n+1} - (\phi S_w \rho_w)^n \right] = 0
\end{aligned}
\tag{7.3.9}
$$

7.3.1.5. Saturation Equation

$$R_{2N_c+2} = 1 - S_o - S_g - S_w = 1 - F \left(\frac{L}{\rho_o} + \frac{1-L}{\rho_g} \right) - S_w = 0$$

where

$$F = \rho_o S_o + \rho_g S_g;$$

$$L = \frac{\rho_o S_o}{F}$$

Difference in the above $2N_c + 2$ equations, the algebraic equations with $2N_c + 2$ unknown variables are established, the solution variables are x_1, $x_2, \ldots, x_{N_c-1}, L, S_w, z_1, z_2, \ldots, z_{N_c-1}, F$ and p.

Algebraic equations are written as follows:

$$\vec{J}\, \delta \vec{x} = -\vec{R}$$

where $\vec{R}$ is remainder term; $\delta \vec{x}$ is the difference in solution variables between two times iteration; $\vec{J}$ is the Jacobian matrix.

$$
\vec{J} =
\begin{bmatrix}
\frac{\partial R_1}{\partial x_1} & \frac{\partial R_1}{\partial x_2} & \cdots & \frac{\partial R_1}{\partial x_{N_c-1}} & \frac{\partial R_1}{\partial L} & \frac{\partial R_1}{\partial S_w} & \frac{\partial R_1}{\partial z_1} & \frac{\partial R_1}{\partial z_2} & \cdots & \frac{\partial R_1}{\partial z_{N_c-1}} & \frac{\partial R_1}{\partial F} & \frac{\partial R_1}{\partial p} \\
\frac{\partial R_2}{\partial x_1} & \frac{\partial R_2}{\partial x_2} & \cdots & \frac{\partial R_2}{\partial x_{N_c-1}} & \frac{\partial R_2}{\partial L} & \frac{\partial R_2}{\partial S_w} & \frac{\partial R_2}{\partial z_1} & \frac{\partial R_2}{\partial z_2} & \cdots & \frac{\partial R_2}{\partial z_{N_c-1}} & \frac{\partial R_2}{\partial F} & \frac{\partial R_2}{\partial p} \\
\vdots & \vdots & & \vdots & \vdots & \vdots & \vdots & \vdots & & \vdots & \vdots & \vdots \\
\frac{\partial R_{2N_c+2}}{\partial x_1} & \frac{\partial R_{2N_c+2}}{\partial x_2} & \cdots & \frac{\partial R_{2N_c+2}}{\partial x_{N_c-1}} & \frac{\partial R_{2N_c+2}}{\partial L} & \frac{\partial R_{2N_c+2}}{\partial S_w} & \frac{\partial R_{2N_c+2}}{\partial z_1} & \frac{\partial R_{2N_c+2}}{\partial z_2} & & \frac{\partial R_{2N_c+2}}{\partial z_{N_c-1}} & \frac{\partial R_{2N_c+2}}{\partial F} & \frac{\partial R_{2N_c+2}}{\partial p}
\end{bmatrix}
$$

$$
\delta \vec{x} =
\begin{bmatrix}
\delta x_i \\
\delta L \\
\delta S_w \\
\delta z_i \\
\delta F \\
\delta p
\end{bmatrix}
\qquad
\vec{R} =
\begin{bmatrix}
R_1 \\
R_1 \\
\vdots \\
\vdots \\
\vdots \\
R_{2N_c+2}
\end{bmatrix}
$$

Because y_i is natural variable, the derivation of y_i must be transformed to the derivation of solution variables z_i, L and x_i.

While $z_i = Lx_i + (1-L)y_i$ and $y_i = \frac{z_i - Lx_i}{1-L}$, the transformation formula is described as follows:

$$\frac{\partial y_i}{\partial z_i} = \frac{1}{1-L}$$

$$\frac{\partial y_i}{\partial L} = \frac{x_i - y_i}{L-1}$$

$$\frac{\partial y_i}{\partial x_i} = \frac{L}{L-1}.$$

7.3.2. Expansion of Cumulative Terms
7.3.2.1. Fugacity Equation

$$R_i = \psi_i^V y_i - \psi_i^L x_i = 0 \qquad (i = 1, 2, \cdots, N_c)$$

$$\frac{\partial R_i}{\partial x_j} = \frac{L}{L-1}\left(\frac{\partial y_i}{\partial y_j}\psi_i^V + y_i\frac{\partial \psi_i^V}{\partial y_j}\right) - \left(\frac{\partial x_i}{\partial x_j}\psi_i^L + x_i\frac{\partial \psi_i^L}{\partial x_j}\right)$$

$$\frac{\partial R_i}{\partial L} = \frac{L}{L-1}\sum_{j=1}^{N_c}(x_j - y_j)\left(\frac{\partial y_i}{\partial y_j}\psi_i^V + y_i\frac{\partial \psi_i^V}{\partial y_j}\right)$$

$$\frac{\partial R_i}{\partial S_w} = 0$$

$$\frac{\partial R_i}{\partial z_j} = \frac{L}{L-1}\left(\frac{\partial y_i}{\partial y_j}\psi_i^V + y_i\frac{\partial \psi_i^V}{\partial y_j}\right)$$

$$\frac{\partial R_i}{\partial F} = 0$$

$$\frac{\partial R_i}{\partial p} = \left(y_i\frac{\partial \psi_i^V}{\partial p} - x_i\frac{\partial \psi_i^L}{\partial p}\right)$$

7.3.2.2. Cumulative Terms of Compositional Flow Equation

$$R_{N_c+i} = -\frac{V}{\Delta t}\left[(\phi F z_i)^{n+1} - (\phi F z_i)^n\right]$$

$$\frac{\partial R_{N_c+i}}{\partial z_j} = -\frac{V_b}{\Delta t}(\phi F)\frac{\partial z_i}{\partial z_j}$$

$$\frac{\partial R_{N_c+i}}{\partial F} = -\frac{V_b}{\Delta t}(\phi z_j)$$

$$\frac{\partial R_{N_c+i}}{\partial p} = -\frac{V_b}{\Delta t}(F z_j)\phi^* C_r$$

$$\frac{\partial R_{N_c+i}}{\partial L} = 0$$

$$\frac{\partial R_{N_c+i}}{\partial S_w} = 0$$

7.3.2.3. Cumulative Terms of Total Hydrocarbon Flow Equation

$$R_{2N_c} = -\frac{V}{\Delta t}(\phi F)$$

$$\frac{\partial R_{2N_c}}{\partial F} = -\frac{V}{\Delta t}\phi$$

$$\frac{\partial R_{2N_c}}{\partial p} = -\frac{V}{\Delta t}F\phi C_r$$

$$\frac{\partial R_{2N_c}}{\partial x_j} = 0$$

$$\frac{\partial R_{2N_c}}{\partial L} = 0$$

$$\frac{\partial R_{2N_c}}{\partial z_j} = 0$$

$$\frac{\partial R_{2N_c}}{\partial S_w} = 0$$

7.3.2.4. Cumulative Terms of Water Flow Equation

$$R_{2N_c+1} = -\frac{V}{\Delta t}(\phi \rho_w S_w)$$

$$\frac{\partial R_{2N_c+1}}{\partial S_w} = -\frac{V}{\Delta t}(\phi \rho_w)$$

$$\frac{\partial R_{2N_c+1}}{\partial p} = -\frac{V}{\Delta t} S_w(\phi^* \rho_w C_w + \phi \rho_w^* C_w)$$

$$\frac{\partial R_{2N_c+1}}{\partial x_j} = 0$$

$$\frac{\partial R_{2N_c+1}}{\partial L} = 0$$

$$\frac{\partial R_{2N_c+1}}{\partial z_j} = 0$$

$$\frac{\partial R_{2N_c+1}}{\partial F} = 0$$

7.3.2.5. Saturation Equation

$$R_{2N_c+2} = 1 - F\left(\frac{L}{\rho_o} + \frac{1-L}{\rho_g}\right) - S_w = 0$$

$$\rho_o = \frac{p}{z_o RT}$$

$$\rho_g = \frac{p}{z_g RT}$$

$$\frac{\partial R_{2N_c+2}}{\partial x_j} = -\frac{FLRT}{p}\left(\frac{\partial z_o}{\partial x_j} - \frac{\partial z_g}{\partial y_j}\right)$$

$$\frac{\partial R_{2N_c+2}}{\partial L} = -\frac{FLRT}{p} - \left[(z_o - z_g) + \sum_{j=1}^{N_c}(y_j - x_j)\frac{\partial z_g}{\partial y_j}\right]$$

$$\frac{\partial R_{2N_c+2}}{\partial S_w} = -1$$

$$\frac{\partial R_{2N_c+2}}{\partial z_j} = \frac{FLRT}{p}\frac{\partial z_g}{\partial y_j}$$

$$\frac{\partial R_{2N_c+2}}{\partial F} = -\frac{RT}{p}\left[Lz_o + (1-L)z_g\right]$$

$$\frac{\partial R_{2N_c+2}}{\partial p} = -\frac{FRT}{p}\left(L\left(\frac{\partial z_o}{\partial p} - \frac{z_o}{p}\right) + (1-L)\left(\frac{\partial z_o}{\partial p} - \frac{z_g}{p}\right)\right)$$

7.3.3. Production Rate Equation

In reservoir conditions, the full rate of one well is shown as follows:

$$q = \sum_p q_p \qquad (7.3.10)$$

$$q_p = q_{lp} + q_{Vp} \qquad (7.3.11)$$

where p represents the perforated layer.

The composition of well fluids:

$$W_{z_i} = \frac{\sum\left(q_{Lp}x_i + q_{Vp}y_i\right)}{q_p} \qquad (7.3.12)$$

In surface conditions, the production rates of oil, gas and water are expressed separately:

$$q_o = q_{Lp}/B_o$$

$$q_g = q_{Vp}/B_g$$

$$q_w = q_w/B_w$$

$$q_{Lp} = WI(kh)_p \lambda_{Lp}\left(p_p - p_{wfp}\right)$$

$$q_{Vp} = WI(kh)_p \lambda_{Vp}\left(p_p - p_{wfp}\right)$$

where WI is well completeness factor; p_p is grid pressure; p_{wfp} is bottom-hole pressure of layer p; p_{wfp}^* is bottom-hole pressure of base level.

$$p_{wfp} = p_{wfp}^* + \gamma(D_p - D^*)$$

$$\lambda_{ol} = \left(\frac{K_{ro}}{\mu_{op}}\right)\rho_{op}\ \lambda_{gl} = \left(\frac{K_{rg}}{\mu_{gp}}\right)\rho_{gp}$$

and then:

$$q_o = WI\left[\sum_p kh_p p_p \left(\lambda_{op} + \lambda_{gl}\right) - \sum_p kh_p p_p \left(\lambda_{op} + \lambda_{gl}\right)\left(p^*_{wfp} + \gamma(D_p - D^*)\right)\right]/B_o$$

$$p^*_{wfp} = \frac{\sum_p kh_p p_p \left(\lambda_{op} + \lambda_{gl}\right)\left(p_p - \gamma(D_p - D^*)\right) - q_o B_o/WI}{\sum_p kh_p \left(\lambda_{op} + \lambda_{gl}\right)}$$

For a production well, if $p^*_{wfp} < (p^*_{wfp})_{min}$ then $p^*_{wfp} = (p^*_{wfp})_{min}$; for an injection well, if $p^*_{wfp} > (p^*_{wfp})_{max}$ then $p^*_{wfp} = (p^*_{wfp})_{max}$.

7.4. STREAMLINE WELL TESTING INTERPRETATION MODEL CONSIDERING COMPONENTS

The pressure of each time-step can be obtained using the above method, and the distribution of pressure, saturation and components along the streamline are obtained using the streamline calculation method in Chapter 2; then transform the distribution of saturation and pressure along the streamline to grids, calculate until testing well shut-in and obtain the distribution of saturation, pressure, components and streamline in the layer before testing. The variation of bottom-hole pressure in the testing well is simulated based on the data.

There are many of streamlines around the well, and the well testing interpretation model is established along each streamline. In this way, three-dimensional well testing problems are simplified to one dimension, which improves the calculation velocity and solution stability. In the well testing interpretation model, well bore storage effect and skin effect of the testing well are considered. Streamline well testing interpretation model considering components are built along each streamline of the testing well, and the model is composed with the filtration equation, inner/outer boundary conditions and initial conditions.

The water phase filtration equation along the streamline is obtained from the water phase mass conservation equation, the Darcy equation and the state equation:

$$\frac{\partial}{\partial l}\left[B\frac{KK_{rw}}{\mu_w}\frac{\partial p_j}{\partial l}\right] = \frac{\partial}{\partial t}(\phi s_w) \qquad (7.4.1)$$

Filtration equation of component i along streamline:

$$\frac{\partial}{\partial l}\left[\left(B\rho_o\frac{KK_{ro}}{\mu_o}x_i\right)\frac{\partial p_j}{\partial l} + \left(B\rho_g\frac{KK_{rg}}{\mu_g}y_i\right)\frac{\partial p_j}{\partial l}\right] = \frac{\partial}{\partial t}\left[\phi\left(\rho_o s_o x_i + \rho_g s_g y_i\right)\right]$$

$$(7.4.2)$$

$$\frac{\partial}{\partial l}\left[\left(B\rho_o\frac{KK_{ro}}{\mu_o}\right)\frac{\partial p_j}{\partial l} + \left(B\rho_g\frac{KK_{rg}}{\mu_g}\right)\frac{\partial p_j}{\partial l}\right] = \frac{\partial}{\partial t}\left[\phi\left(\rho_o s_o + \rho_g s_g\right)\right] \qquad (7.4.3)$$

where B is the cross-sectional area of the flow tube, which is different in every grid.

Initial condition: the distribution of saturation and pressure along each streamline are known, which is obtained from black oil model.

For inner boundary conditions, two situations (production well and injection well) are included:

1. Inner boundary condition of the production well.

$$\left[\left(\rho_o \frac{B_j KK_{ro}}{\mu_o} + \rho_w \frac{B_j KK_{rw}}{\mu_w} + \rho_g \frac{B_j KK_{rg}}{\mu_g}\right) \frac{\partial p_j}{\partial l}\right]_{l=r_w} - \frac{24C}{N} \frac{dp_j}{dt} = 0$$

The above formula can be simplified as follows:

$$\left(B_i \lambda_t \frac{\partial p_j}{\partial l}\right)_{l=r_w} - \frac{24C}{N} \frac{dp_{wj}}{dt} = \frac{\theta_j}{2\pi} \sum q \tag{7.4.4}$$

$$p_{wj} = p_{j1} + \Delta p_{sj} = p_{j1} + s \frac{\theta_j}{2\pi} \left(\frac{\partial p_j}{\partial l}\right)_{l=r_w} \tag{7.4.5}$$

$$p_w = \frac{1}{N} \sum_{j=1}^{N} p_{wj} \tag{7.4.6}$$

2. Inner boundary condition of the injection well.

For the inner boundary condition of the injection well, two situations are included: one is the water injection well, the other is the gas injection well.

   i. Water injection well.

$$\left[\left(\frac{B_j KK_{rw}}{\mu_w}\right) \frac{\partial p_j}{\partial l}\right]_{l=r_w} - \frac{24C}{N_j} \frac{dp_j}{dt} = \frac{\theta_j}{2\pi} q_w$$

The above formula can be simplified as follows:

$$\left(B_j \lambda_w \frac{\partial p_j}{\partial l}\right)_{l=r_w} - \frac{24C}{N} \frac{dp_{wj}}{dt} = \frac{\theta_j}{2\pi} q_w \tag{7.4.7}$$

$$p_{wj} = p_{j1} + \Delta p_{sj} = p_{j1} + s \frac{\theta_j}{2\pi} \left(\frac{\partial p_j}{\partial l}\right)_{l=r_w} \tag{7.4.8}$$

$$p_w = \frac{1}{N} \sum_{j=1}^{N} p_{wj} \tag{7.4.9}$$

   ii. Gas injection well.

$$\left[\left(\frac{B_j KK_{rg}}{\mu_g}\right) \frac{\partial p_j}{\partial l}\right]_{l=r_w} - \frac{24C}{N} \frac{dp_{wj}}{dt} = \frac{\theta_j}{2\pi} q_g$$

The above formula can be simplified as follows:

$$\left(B_i\lambda_g\frac{\partial p_j}{\partial l}\right)_{l=r_w} - \frac{24C}{N}\frac{dp_{wj}}{dt} = \frac{\theta_j}{2\pi}q_g \qquad (7.4.10)$$

$$p_{wj} = p_{j1} + \Delta p_{sj} = p_{j1} + S\frac{\theta_j}{2\pi}\left(\frac{\partial p_j}{\partial l}\right)_{l=r_w} \qquad (7.4.11)$$

$$p_w = \frac{1}{N}\sum_{j=1}^{N}p_{wj} \qquad (7.4.12)$$

Where N is the number of streamlines around the testing well; θ_j is the angle of the jth streamline; p_{j1} is the pressure of the jth streamline on the well wall; C is well bore storage factor; p_w is bottom-hole pressure.

7.5. DISCRETE OF STREAMLINE WELL TESTING INTERPRETATION MODEL CONSIDERING COMPONENTS

In this section, the IMPES method of reservoir numerical simulation is used to solve the compositional well testing interpretation model which is built in section 2.

7.5.1. Difference of Well Testing Interpretation Model
7.5.1.1. Differential Discrete of Water Phase Equation

Difference of water phase flow Equation 7.4.1:

$$\frac{\left[B\frac{KK_{rw}}{\mu_w}\frac{\partial p_j}{\partial l}\right]_{i+\frac{1}{2}} - \left[B\frac{KK_{rw}}{\mu_w}\frac{\partial p_j}{\partial l}\right]_{i-\frac{1}{2}}}{l_{i+\frac{1}{2}} - l_{i-\frac{1}{2}}} = \frac{\phi}{\Delta t}\left(s_{wi}^{n+1} - s_{wi}^{n}\right)$$

$$\frac{\left[(BK)_{i+\frac{1}{2}}\left(\frac{K_{rw}}{\mu_w}\right)_{i+\frac{1}{2}}\frac{p_{ji+1}^{n+1}-p_{ji}^{n+1}}{l_{i+1}-l_i} - (BK)_{i-\frac{1}{2}}\left(\frac{K_{rw}}{\mu_w}\right)_{i-\frac{1}{2}}\frac{p_{ji}^{n+1}-p_{ji-1}^{n+1}}{l_i-l_{i-1}}\right]}{l_{j+\frac{1}{2}} - l_{j-\frac{1}{2}}} = \frac{\phi}{\Delta t}\left(s_{wi}^{n+1} - s_{wi}^{n}\right)$$

Assume:

$$A_{w1} = \frac{(BK)_{i+\frac{1}{2}}\left(\frac{K_{rw}}{\mu_w}\right)_{i+\frac{1}{2}}}{\left(l_{i+\frac{1}{2}} - l_{i-\frac{1}{2}}\right)\left(l_{i+1} - l_i\right)}$$

$$A_{w2} = \frac{(BK)_{i-\frac{1}{2}}\left(\frac{K_{rw}}{\mu_w}\right)_{i-\frac{1}{2}}}{\left(l_{i+\frac{1}{2}} - l_{i-\frac{1}{2}}\right)\left(l_{i+1} - l_i\right)}$$

Then

$$A_{w1}\left(p_{ji+1}^{n+1} - p_{ji}^{n+1}\right) + A_{w2}\left(p_{ji}^{n+1} - p_{ji-1}^{n+1}\right) = \frac{\phi}{\Delta t}\left(s_{wi}^{n+1} - s_{wi}^{n}\right) \qquad (7.5.1)$$

7.5.1.2. Discretization of Total Hydrocarbon Flow Equation

Difference of Equation 7.4.3:

$$
\frac{\left[BK\left(\rho_o\dfrac{K_{ro}}{\mu_o}\dfrac{\partial p_j}{\partial l}+\rho_g\dfrac{K_{rg}}{\mu_g}\dfrac{\partial p_j}{\partial l}\right)\right]_{i+\frac{1}{2}}-\left[BK\left(\rho_o\dfrac{K_{ro}}{\mu_o}\dfrac{\partial p_j}{\partial l}+\rho_g\dfrac{K_{rg}}{\mu_g}\dfrac{\partial p_j}{\partial l}\right)\right]_{i-\frac{1}{2}}}{l_{i+\frac{1}{2}}-l_{i-\frac{1}{2}}}
$$

$$
=\frac{\phi}{\Delta t}\left[\left(\rho_o s_o+\rho_g s_g\right)_i^{n+1}-\left(\rho_o s_o+\rho_g s_g\right)_i^{n}\right]
$$

$$
\frac{1}{\left(l_{i+\frac{1}{2}}-l_{i-\frac{1}{2}}\right)}\left[(BK)_{i+\frac{1}{2}}\left(\rho_o\frac{K_{ro}}{\mu_o}\right)_{i+\frac{1}{2}}\frac{p_{ji+1}^{n+1}-P_{ji}^{n+1}}{l_{i+1}-l_i}-(BK)_{i-\frac{1}{2}}\left(\rho_o\frac{K_{ro}}{\mu_o}\right)_{i-\frac{1}{2}}\frac{p_{ji}^{n+1}-P_{ji-1}^{n+1}}{l_i-l_{i-1}}\right.
$$

$$
\left.+(BK)_{i+\frac{1}{2}}\left(\rho_g\frac{K_{rg}}{\mu_g}\right)_{i+\frac{1}{2}}\frac{p_{ji+1}^{n+1}-P_{ji}^{n+1}}{l_{i+1}-l_i}-(BK)_{i-\frac{1}{2}}\left(\rho_g\frac{K_{rg}}{\mu_g}\right)_{i-\frac{1}{2}}\frac{p_{ji}^{n+1}-P_{ji-1}^{n+1}}{l_i-l_{i-1}}\right]
$$

Assume:

$$
A_{o1}=\frac{(BK)_{i+\frac{1}{2}}\left(\rho_o\frac{K_{ro}}{\mu_o}\right)_{i+\frac{1}{2}}}{\left(l_{i+\frac{1}{2}}-l_{i-\frac{1}{2}}\right)\left(l_{i+1}-l_i\right)}
$$

$$
A_{o2}=\frac{(BK)_{i-\frac{1}{2}}\left(\rho_o\frac{K_{ro}}{\mu_o}\right)_{i-\frac{1}{2}}}{\left(l_{i+\frac{1}{2}}-l_{i-\frac{1}{2}}\right)\left(l_{i+1}-l_i\right)}
$$

$$
A_{g1}=\frac{(BK)_{i+\frac{1}{2}}\left(\rho_g\frac{K_{rg}}{\mu_g}\right)_{i+\frac{1}{2}}}{\left(l_{i+\frac{1}{2}}-l_{i-\frac{1}{2}}\right)\left(l_{i+1}-l_i\right)}
$$

$$
A_{g2}=\frac{(BK)_{i-\frac{1}{2}}\left(\rho_g\frac{K_{rg}}{\mu_g}\right)_{i-\frac{1}{2}}}{\left(l_{i+\frac{1}{2}}-l_{i-\frac{1}{2}}\right)\left(l_{i+1}-l_i\right)}
$$

Then

$$
A_{o1}\left(p_{ji+1}^{n+1}-p_{ji}^{n+1}\right)+A_{o2}\left(p_{ji}^{n+1}-p_{ji-1}^{n+1}\right)+A_{g1}\left(p_{ji+1}^{n+1}-p_{ji}^{n+1}\right)+A_{g2}\left(p_{ji}^{n+1}-p_{ji-1}^{n+1}\right)
$$

$$
=\frac{\phi}{\Delta t}\left[\left(\rho_o s_o+\rho_g s_g\right)_i^{n+1}-\left(\rho_o s_o+\rho_g s_g\right)_i^{n}\right]
$$

$$
(7.5.2)
$$

7.5.1.3. Pressure Equation

Multiply $\beta = \left(\frac{s_o + s_g}{\rho_o s_o + \rho_g s_g}\right)^{n+1}$ with Equation 7.5.2, and add Equation 7.5.1 with Equation 7.5.2, then

The left side $= \left(A_{w1} + \beta A_{o1} + \beta A_{g1}\right)p_{ji+1}^{n+1} - (A_{w1} - A_{w2} + \beta A_{o1} - \beta A_{o2}$
$+ \beta A_{g1} - \beta A_{g2})p_{ji}^{n+1} - (A_{w2} + \beta A_{o2} + \beta A_{g2})p_{ji-1}^{n+1}$

The right side

$$= \frac{\phi}{\Delta t}\left[s_{wi}^{n+1} - s_{wi}^{n}\right] + \frac{\phi}{\Delta t}\left[\left(\rho_o s_o + \rho_g s_g\right)_i^{n+1} - \left(\rho_o s_o + \rho_g s_g\right)_i^{n}\right] \cdot \left(\frac{s_o + s_g}{\rho_o s_o + \rho_g s_g}\right)_i^{n+1}$$

$$= \frac{\phi}{\Delta t}\left[s_{wi}^{n+1} - s_{wi}^{n}\right] + \frac{\phi}{\Delta t}\left[\left(s_o + s_g\right)_i^{n+1} - \beta\left(\rho_o s_o + \rho_g s_g\right)_i^{n}\right]$$

$$= \frac{\phi}{\Delta t}\left[s_{wi}^{n+1} + s_{oi}^{n+1} + s_g^{n+1} - s_{wi}^{n} - \beta\left(\rho_o s_o + \rho_g s_g\right)_i^{n}\right]$$

$$= \frac{\phi}{\Delta t}\left[1 - s_{wi}^{n} - \beta\left(\rho_o s_o + \rho_g s_g\right)_i^{n}\right]$$

Assume:

$$d_i = A_{w1} + \beta A_{o1} + \beta A_{g1}$$

$$c_i = -\left(A_{w1} - A_{w2} + \beta A_{o1} - \beta A_{o12} + \beta A_{g1} - \beta A_{g2}\right)$$

$$b_i = -(A_{w2} + \beta A_{o2} + \beta A_{g2})$$

$$g_i = \frac{\phi}{\Delta t}\left[1 - s_{wi}^{n} - \beta\left(\rho_o s_o + \rho_g s_g\right)_i^{n}\right]$$

Then

$$b_i p_{i-1}^{n+1} + c_i p_i^{n+1} + d_i p_{i+1}^{n+1} = g_i \qquad 2 \le i \le N_j - 1 \qquad (7.5.3)$$

7.5.2. Difference in Inner Boundary Condition
7.5.2.1. Production Well

Discretization of Equations 7.4.4 and 7.4.5:

$$\left[\left(B\rho_o \frac{kk_{ro}}{\mu_o}\right)_1 \frac{p_{j2}^{n+1} - p_{j1}^{n+1}}{l_2 - l_1} + \left(B\rho_g \frac{kk_{rg}}{\mu_g}\right)_1 \frac{p_{j2}^{n+1} - p_{j1}^{n+1}}{l_2 - l_1}\right]$$
$$+ \left[B\frac{kk_{rw}}{\mu_w}\right]_1 \frac{p_{j2}^{n+1} - p_{j1}^{n+1}}{l_2 - l_1} - \frac{24C}{N}\frac{p_{wj}^{n+1} - p_{wj}^{n}}{\Delta t} = 0$$

$$p_{wj}^{n+1} = p_j^{n+1} + \frac{S\left(p_{j2}^{n+1} - p_{j1}^{n+1}\right)}{l_2 - l_1}$$

Assume:

$$A_{co} = \frac{\left[B\rho_o \frac{kk_{ro}}{\mu_o} \right]_1}{l_2 - l_1}$$

$$A_{cg} = \frac{\left[B\rho_g \frac{kk_{rg}}{\mu_g} \right]_1}{l_2 - l_1}$$

$$A_{cw} = \frac{\left[B \frac{kk_{rw}}{\mu_w} \right]_1}{l_2 - l_1}$$

$$c_1 = -A_{co} - A_{cg} - A_{cw} - \frac{24C}{N\Delta t} + \frac{24CS}{N\Delta t(l_2 - l_1)}$$

$$d_1 = A_{co} + A_{cg} + A_{cw} - \frac{24CS}{N\Delta t(l_2 - l_1)}$$

$$g_1 = \frac{\theta_j}{2\pi} \sum q - \frac{24C}{N\Delta t}$$

Then

$$c_1 p_{j1}^{n+1} + d_1 p_{j2}^{n+1} = g_1 \qquad i = 1 \qquad (7.5.4a)$$

7.5.2.2. Injection Well
7.5.2.2.1. Water Injection Well
Discretization of Equations 7.4.7 and 7.4.8:

$$\left[B\frac{kk_{rw}}{\mu_w} \right]_1 \frac{p_{j2}^{n+1} - p_{j1}^{n+1}}{l_2 - l_1} - \frac{24C}{N} \frac{p_{wj}^{n+1} - p_{wj}^n}{\Delta t} = 0$$

$$p_{wj}^{n+1} = p_j^{n+1} + \frac{S\left(p_{j2}^{n+1} - p_{j1}^{n+1} \right)}{l_2 - l_1}$$

Assume:

$$A_{cg} = \frac{\left[B\rho_g \frac{kk_{rg}}{\mu_g} \right]_1}{l_2 - l_1}$$

$$c_1 = -A_{cw} - \frac{24C}{N\Delta t} + \frac{24CS}{N\Delta t(l_2 - l_1)}$$

$$d_1 = A_{cw} - \frac{24CS}{N\Delta t(l_2 - l_1)}$$

$$g_1 = \frac{\theta_j}{2\pi}q_w - \frac{24C}{N\Delta t}$$

Then

$$c_1 p_{j1}^{n+1} + d_1 p_{j2}^{n+1} = g_1 \qquad (i = 1) \qquad (7.5.4b)$$

7.5.2.2.2. Gas Injection Well

$$\left(\rho_g \frac{BKK_{rg}}{\mu_g}\right)_1 \frac{p_{j2}^{n+1} - p_{j1}^{n+1}}{l_2 - l_1} - \frac{24C}{N}\frac{p_{wj}^{n+1} - p_{wj}^n}{\Delta t} = 0$$

$$p_{wj}^{n+1} = p_j^{n+1} + \frac{S\left(p_{j2}^{n+1} - p_{j1}^{n+1}\right)}{l_2 - l_1}$$

Assume:

$$A_{cg} = \frac{\left[B\rho_g \frac{KK_{rg}}{\mu_g}\right]_1}{l_2 - l_1}$$

$$c_1 = -A_{cg} - \frac{24C}{N\Delta t} + \frac{24CS}{N\Delta t(l_2 - l_1)}$$

$$d_1 = A_{cg} - \frac{24CS}{N\Delta t(l_2 - l_1)}$$

$$g_1 = \frac{\theta_j}{2\pi}q_g - \frac{24C}{N\Delta t}$$

Then

$$c_1 p_{j1}^{n+1} + d_1 p_{j2}^{n+1} = g_1 \qquad (i = 1) \qquad (7.5.4c)$$

The three inner boundary conditions above could be written in a unified form.

7.5.3. Difference in Outer Boundary Condition
7.5.3.1. If Streamline Outer Boundary Condition Connects with Well, Two Situations are Included

1. Keep well pressure as a constant during testing, namely constant bottom-hole flowing pressure
That is,

$$p_{jN_j}^{n+1} = p_{wf}$$

In the form of a tridiagonal equation:

$$c_{N_j} = 1, b_{N_j} = 0, g_{N_j} = p_{wf}$$

$$b_{N_j} p_{jN_j-1}^{n+1} + c_{N_j} p_{jN_j}^{n+1} = g_{N_j} \qquad (7.5.5a)$$

2. Keep well production rate as a constant during testing, namely constant production rate

For the water injection well:

$$q_w = -\frac{KK_{rw}hB_j}{\mu_w}\frac{\partial p}{\partial l}\bigg|_{N_j}$$

Where B_j is the streamline width of the outer well.

Difference in the above formula and change into the form of tridiagonal equation:

$$b_{N_j}p_{jN_{j-1}}^{n+1} + c_{N_j}p_{jN_j}^{n+1} = g_{N_j} \tag{7.5.5b}$$

where

$$b_{N_j} = 1$$

$$c_{N_j} = -1$$

$$g_{N_j} = \frac{q_w\mu_w}{KK_{rw}B_j}\left(l_{N_j} - l_{N_{j-1}}\right)$$

For the production well:

$$q_o + q_g + q_w = \left(\frac{KK_{ro}B_j}{\mu_o} + \frac{KK_{rg}B_j}{\mu_g} + \frac{KK_{rw}B_j}{\mu_w}\right)\frac{\partial p_j}{\partial l}$$

That is:

$$q = \left(B_j\lambda_t\frac{\partial p_j}{\partial l}\right)$$

Discrete the above formula:

$$b_{N_j}p_{jN_{j-1}}^{n+1} + c_{N_j}p_{jN_j}^{n+1} = g_{N_j} \tag{7.5.5c}$$

where

$$b_{N_j} = 1$$

$$c_{N_j} = -1$$

$$g_{N_j} = \frac{q_w}{\lambda_{tN_j}hB_j}\left(l_{N_j} - l_{N_{j-1}}\right)$$

7.5.3.2. Scenario of Outer Boundary Divided into Constant Pressure Boundary and Closed Boundary

For the constant pressure boundary, the equation is the same as Equation 7.5.5a of constant bottom-hole flowing pressure.

For closed boundary:

$$\frac{\partial p_j}{\partial l}\Big|_{N_j} = 0$$

$$\frac{\partial p_j}{\partial l}|N_j = 0$$

Difference:

$$p_{jN_j-1}^{n+1} - p_{jN_j}^{n+1} = 0$$

where

$$C_{N_j} = -1$$

$$b_{N_j} = 1$$

$$g_{N_j} = 0$$

$$b_{N_j}p_{jN_{j-1}}^{n+1} + C_{N_j}p_{jN_j}^{n+1} = g_{N_j}. \qquad (7.5.5d)$$

The two outer boundary conditions above could be written in unified form.

7.5.4. Matrix Form of Equations

The matrix form can be written from Equations 7.5.3, 7.5.4 and 7.5.5:

$$\begin{bmatrix} c_1 & d_1 & & & & & \\ b_2 & c_2 & d_2 & & & & \\ & b_3 & c_3 & d_3 & & & \\ & & \ddots & \ddots & \ddots & & \\ & & & \ddots & \ddots & \ddots & \\ & & & & \ddots & \ddots & \ddots \\ & & & & & b_{N_j-1} & c_{N_j-1} & d_{N_j-1} \\ & & & & & & b_{N_j} & c_{N_j} \end{bmatrix} \begin{bmatrix} p_{j1}^{n+1} \\ p_{j2}^{n+1} \\ p_{j3}^{n+1} \\ \vdots \\ \vdots \\ \vdots \\ p_{jN_j-1}^{n+1} \\ p_{jN_j}^{n+1} \end{bmatrix} = \begin{bmatrix} g_1 \\ g_2 \\ g_3 \\ \vdots \\ \vdots \\ \vdots \\ g_{N_j-1} \\ g_{N_j} \end{bmatrix} \qquad (7.5.6)$$

The above equation is a tridiagonal equation, and the chasing method is used to solve it. Calculate all the streamlines using the above method and obtain the bottom-hole pressure:

$$p_{wj}^{n+1} = p_{w1}^{n+1} + S\frac{\theta_j}{2\pi}\frac{p_{j2}^{n+1} - p_{j1}^{n+1}}{l_2 - l_1}$$

Hence, the bottom-hole pressure of the testing well can be established as follows:

$$p_w^{n+1} = \frac{1}{N} \sum_{j=1}^{N} p_{wj}^{n+1}$$

Repeat the above process and obtain the theoretical pressure response.

7.6. COMPONENT AND SATURATION CALCULATION ALONG STREAMLINE

After finishing the pressure calculation of each step, the component and saturation along streamline can be calculated using the following method.

7.6.1. Component Calculation Along Streamline

With the definition of component parameter:

$$k_i = y_i/x_i \tag{7.6.1}$$

$$z_i = y_i V + x_i L. \tag{7.6.2}$$

$$x_i = \frac{z_i}{1 + V(k_i - 1)} \tag{7.6.3}$$

$$y_i = \frac{k_i z_i}{1 + V(k_i - 1)}. \tag{7.6.4}$$

$$L + V = 1 \tag{7.6.5}$$

$$V = \frac{\rho_g S_g}{\rho_g S_g + \rho_o S_o} \tag{7.6.6}$$

Replace the right side of Equations 7.4.2 with 7.6.3, 7.6.4 and 7.6.6, then:

$$\text{Right side} = \frac{1}{1 + \frac{\rho_g S_g}{\rho_g S_g + \rho_o S_o}(k_i - 1)} (k_i \rho_g S_g + \rho_o S_o) z_i = z_i (\rho_g S_g + \rho_o S_o)$$

$$\tag{7.6.7}$$

Take Equation 5.6.7 into the right side of Equation 5.4.2, make difference:

$$\frac{1}{\left(l_{i+\frac{1}{2}} - l_{i-\frac{1}{2}}\right)} \left[K_{i+\frac{1}{2}} \left(\rho_o x_i \frac{K_{ro}}{\mu_o} \right)^n_{i+\frac{1}{2}} \frac{P_{ji+1}^{n+1} - P_{ji}^{n+1}}{l_{i+1} - l_i} - K_{i-\frac{1}{2}} \left(\rho_o x_i \frac{K_{ro}}{\mu_o} \right)^n_{i-\frac{1}{2}} \frac{P_{ji}^{n+1} - P_{ji-1}^{n+1}}{l_i - l_{i-1}} \right)$$

$$+ K_{i+\frac{1}{2}} \left(\rho_g y_i \frac{K_{rg}}{\mu_g} \right)^n_{i+\frac{1}{2}} \frac{P_{ji+1}^{n+1} - P_{ji}^{n+1}}{l_{i+1} - l_i} - K_{i-\frac{1}{2}} \left(\rho_g y_i \frac{K_{rg}}{\mu_g} \right)^n_{i-\frac{1}{2}} \frac{P_{ji}^{n+1} - P_{ji-1}^{n+1}}{l_i - l_{i-1}}$$

$$= \frac{\phi}{\Delta t} \left[z_i^{n+1} \left(\rho_o S_o + \rho_g S_g \right)^{n+1}_i - z_i^n \left(\rho_o S_o + \rho_g S_g \right)^n_i \right] \tag{7.6.8}$$

Parameter z_m^{n+1} is unknown among all parameters at time $n + 1$, so z_m^{n+1} can be calculated from Equation 7.6.8.

7.6.2. Saturation Calculation Along Streamline

After getting $p_{ji}^{n+1} \left(i = 1, 2, \cdots, N_j \right)$, water saturation at time $n + 1$ is obtained from Equation 7.5.1:

$$S_{wi}^{n+1} = S_{wi}^n + \frac{\Delta t}{\phi} \left[A_{w1} \left(p_{ji+1}^{n+1} - p_{ji}^{n+1} \right) + A_{w2} \left(p_{ji}^{n+1} - p_{ji-1}^{n+1} \right) \right]. \qquad (7.6.9)$$

Phase equilibrium flash calculation is taken to obtain L^{n+1}, V^{n+1} and $[x_i^{n+1}, y_i^{n+1}, i = 1, 2, 3, \ldots, N]$ after the calculation of $p_{ji}^{n+1} \left(i = 1, 2, \cdots, N_j \right)$, $z_{ji}^{n+1} \left(i = 1, 2, \cdots, N_j \right)$ and $S_{wji}^{n+1} \left(i = 1, 2, \cdots, N_j \right)$. At last, with the equilibrium saturation formula:

$$L^{n+1} = S_o^{n+1} \rho_o^{n+1} / \left(S_o^{n+1} \rho_o^{n+1} + S_g^{n+1} \rho_g^{n+1} \right)$$

$$S_g^{n+1} = 1 - S_o^{n+1} - S_w^{n+1}$$

S_g^{n+1} and S_o^{n+1} are established:

$$S_g^{n+1} = \left[\frac{(1 - S_W) \rho_o (1 - L)}{L \rho_g + (1 - L) \rho_o} \right]^{n+1} \qquad (7.6.10)$$

$$S_o^{n+1} = \left[\frac{(1 - S_W) \rho_g L}{L \rho_g + (1 - L) \rho_o} \right]^{n+1} \qquad (7.6.11)$$

7.7. SIMULATION EXAMPLE ANALYSIS

7.7.1. Basic Data
7.7.1.1. Reservoir Fluids, Composition of Injection Fluids and Relevant Parameters

For simplification, pseudo-components are used to represent real fluid components. The contents of six pseudo-components and relevant parameters are shown in Table 7.1.

7.1.2. Basic Parameters of Reservoir and Wells

Reservoir area is 500×500 m, layer thickness $h = 15$ m, porosity $\phi = 0.2$, average formation permeability $K = 200 \times 10^{-3}\,\mu m^2$, initial oil saturation $S_o = 0.76$, irreducible water saturation $S_w = 0.24$, initial formation pressure $p_i = 40$ MPa, oil well damage factor $S = 0$, well bore storage factor $C = 0.15 m^3 / MPa$.

7.7.2. Testing Simulation
7.7.2.1. Pressure Drawdown Well Testing Simulation of Production Well Before the Miscible Phase

For the above reservoir model, with only one production well in the center, production rate $q_o = 40 m^3 / d$, the compositional well testing simulator

TABLE 7.1 Fluid Pseudo-Component, Injected Hydrocarbon Component and Relevant Parameters

Parameter	Component					
	C_1, N_2	C_2, CO_2	C_3, C_4	C_5, C_6	$C_7 \sim C_{16}$	C_{17+}
P_c/MPa	5.636	6.844	5.108	4.764	3.359	2.419
T_c/°C	18.63	135.3	219.3	315.8	460.8	655.5
V_c (m³/mol)	0.098	0.144	0.224	0.335	0.583	1.078
Molecular weight/ (kg/mol)	16.39	30.83	49.61	78.81	147.0	319.4
Eccentric factor	0.008	0.104	0.164	0.265	0.451	0.802
Initial hydrocarbon composition/%	57.94	9.860	10.96	3.980	13.67	3.590
Injected hydrocarbon composition/%	78.70	11.80	8.400	0.600	0.100	0.100

is used to calculate pressure drawdown response, and the pressure curve is shown in Fig. 7.1.

7.7.2.2. Pressure Fall-Off Well Testing Simulation of Gas Injection Well After Miscible Phase

In order to simulate the miscible phase after gas injection in the above reservoir, the gas injection well is located in the center while the surrounding wells produce oil at the same time. Inject gas for 200 days and simulate the pressure drawdown test of the gas injection well in the center, all the wells are shut in

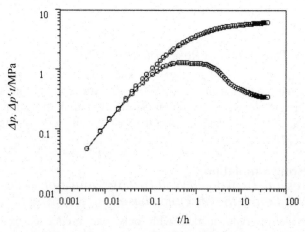

FIGURE 7.1 Pressure drawdown curve of production well before the miscible phase.

before testing. The location of reservoir wells is shown in Fig. 7.2, and the testing curve is shown in Fig. 7.3.

7.7.3. Testing Curve Analysis

For the pressure drawdown test before the miscible phase, because there are no changes in sub-surface fluid components, fluid viscosity, density and compressibility factor, the simulation testing curve shows the characteristics of a homogeneous reservoir.

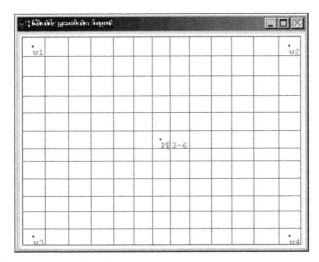

FIGURE 7.2 Reservoir well location.

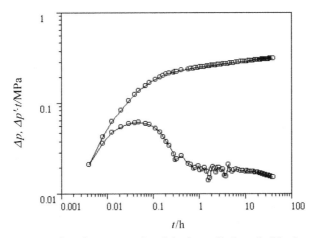

FIGURE 7.3 Pressure drawdown curve of gas injection well after miscible phase.

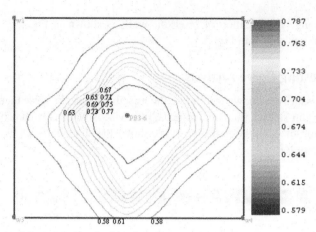

FIGURE 7.4   Sub-surface distribution of the first component (C1, N2).

For the pressure drawdown test after the miscible phase, because of the miscible phase between injection fluid and sub-surface fluid (the minimum miscible phase pressure in Pubei oilfield is 33 MPa), the components of sub-surface fluid change, as do the fluid viscosity, density and compressibility factor. As seen in the pressure derivative curve: compared with that before the miscible phase, derivation falls quickly after well bore storage effect; the main reason is due to viscosity variation of sub-surface fluid after miscible phase. The fluctuation of derivative curve is due to the instability of numerical solutions. The component distribution of sub-surface hydrocarbon at the time of well shut-in is shown in Figs 7.4 to 7.9.

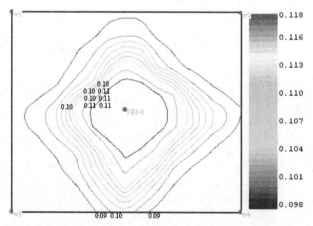

FIGURE 7.5   Sub-surface distribution of the second component (C_2, CO_2).

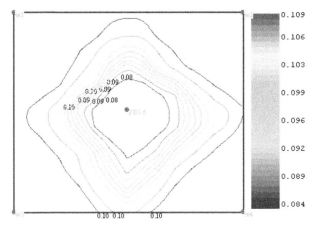

FIGURE 7.6 Sub-surface distribution of the third component (C3, C4).

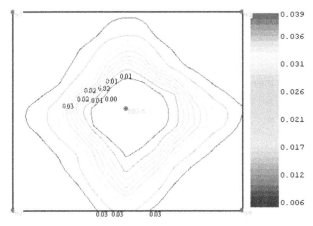

FIGURE 7.7 Sub-surface distribution of the fourth component (C_5, C_6).

7.8. CHAPTER SUMMARY

In this chapter, streamline numerical well testing interpretation method considering components are studied. The main contents include:

1. The streamline numerical well testing interpretation model considering components include the filtration mathematical model of the production period and the streamline mathematical model of the testing period. The filtration mathematical model of the production period is a compositional model, including the water phase flow equation, the oil/gas phase flow equation, the phase equilibrium equation, the auxiliary equation, and inner/outer boundary conditions; the streamline method is used to solve

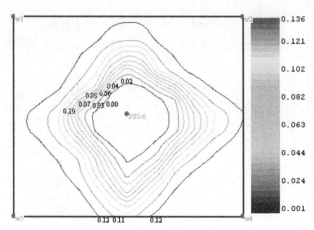

FIGURE 7.8 Sub-surface distribution of the fifth component (C7~C16).

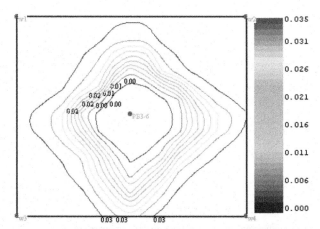

FIGURE 7.9 Sub-surface distribution of the sixth component (C$_{17+}$).

this. The mathematical model of the testing period is the streamline model, including the water and component flow control equation, inner- and outer-boundary conditions; the solving method is finite difference.

2. With the simulation examples of typical and real reservoir models, the characteristics of the streamline numerical well testing interpretation model considering components are studied.

Streamline Numerical Well Testing Interpretation Model for a Multi-Layer Reservoir in Double-Porosity Media

The term "dual-porosity media" means that there are two kinds of porous media with different properties in the reservoir, and the media are combined by block media with pore space and fracture media with a divisional block. They overlap on space, so they are independent of and connected to each other; also, mass transfer happens under certain conditions. The block media, which is named matrix, is continuous with the fracture on space. Usually, the sectional area of fracture is much larger than that of the matrix pore, so the permeability of the fracture system is obviously greater than that of the matrix rock, and it is the main fluid flow channel; meanwhile, the volume of the fracture system is much less than that of the matrix pore, so the porosity of the matrix system is obviously greater than that of the fracture system, and it is the main storage space. Because of the difference in permeability and storage ability in the two media, two parallel flow fields are formed and there are fluid exchanges between them; this is defined as "cross-flow".

8.1. ESTABLISHMENT OF THE STREAMLINE NUMERICAL WELL TESTING INTERPRETATION MODEL FOR A MULTI-LAYER RESERVOIR IN DOUBLE-POROSITY MEDIA

The streamline numerical well testing interpretation model for a multi-layer reservoir in dual-porosity media includes the filtration mathematical model of the production period and the streamline mathematical model of the testing period.

8.1.1. Physical Model

Assumptions:

1. reservoir is in dual porosity media, and includes oil and water phase fluids;
2. rock and fluid are incompressible in the reservoir;

Streamline Numerical Well Test Interpretation. DOI: 10.1016/B978-0-12-386027-9.00008-3
Copyright © 2011 by Elsevier Ltd

3. the permeability of the fracture system is larger than that of the matrix system, and fluid flows between fractures and follows the Darcy law;
4. the porosity of the matrix system is larger than that of the fracture system, and there is no fluid flow between blocks;
5. cross-flow exists between the matrix system and the fracture system; and
6. the effects of gravity and capillary force are ignored.

The physical model of dual-porosity media is shown in Fig. 8.1.

8.1.2. Filtration Mathematical Model of the Production Period
8.1.2.1. Pressure Control Equation
Subscript f represents the fracture system, and m represents the matrix system. For the fracture system, the pressure control equation of oil and water phases is shown, respectively:

$$\nabla \cdot (K_f \lambda_{of} \nabla p_f) + T_o + \frac{q_{of}}{\rho_o} = \phi_f \frac{\partial S_{of}}{\partial t} \qquad (8.1.1)$$

$$\nabla \cdot (K_f \lambda_{wf} \nabla p_f) + T_w + \frac{q_{wf}}{\rho_w} = \phi_f \frac{\partial S_{wf}}{\partial t} \qquad (8.1.2)$$

From saturation normalization equation $S_{of} + S_{wf} = 1$, Equations 8.1.1 and 8.1.2 are combined as:

$$\nabla \cdot (K_f \lambda_{tf} \nabla p_f) + T_t = -q_{sf} \qquad (8.1.3)$$

For the matrix system, the pressure control equation of oil and water phases is shown, respectively:

$$\nabla \cdot (K_m \lambda_{om} \nabla p_m) - T_o + \frac{q_{om}}{\rho_o} - \phi_m \frac{\partial S_{om}}{\partial t} \qquad (8.1.4)$$

$$\nabla \cdot (K_m \lambda_{wm} \nabla p_m) - T_w + \frac{q_{wm}}{\rho_w} - \phi_m \frac{\partial S_{wm}}{\partial t} \qquad (8.1.5)$$

From saturation normalization equation $S_{om} + S_{wm} = 1$, Equations 8.1.4 and 8.1.5 are combined as:

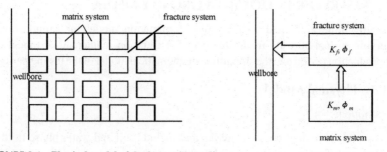

FIGURE 8.1   Physical model of dual-porosity media.

$$\nabla \cdot (K_m \lambda_{tm} \nabla p_m) - T_t = -q_{sm} \qquad (8.1.6)$$

where, T_o, T_w and T_t represent the cross-flow rate of oil phase, water phase and total between matrix and fracture, $T_t = T_o + T_w$, s^{-1}; q_o, q_w and q_s are source/sink term, and $q_s = q_o + q$ [g/(cm$^3 \bullet$ s)].

8.1.2.2. Cross-Flow Function Between Systems

The cross-flow function between systems is used to describe the fluid exchange between blocks and fractures. While the fluid exchange between blocks and fracture happens under slowly changed pressure, this process is considered as pseudo-steady:

$$T_w = F_s K_m \lambda_w (p_f - p_m) \qquad (8.1.7)$$

$$T_o = F_s K_m \lambda_o (p_f - p_m) \qquad (8.1.8)$$

where F_s is the shape factor, which is used to describe cross-flow ability between blocks and fractures (cm^{-2}).

8.1.2.3. Initial Condition

$$\left.\begin{array}{l} p_f|_{t=0} = p_i \\ p_m|_{t=0} = p_i \end{array}\right\} \qquad (8.1.9)$$

$$\left.\begin{array}{l} S_{wf}|_{t=0} = S_{wi} \\ S_{wm}|_{t=0} = S_{wi} \end{array}\right\} \qquad (8.1.10)$$

8.1.2.4. Inner Boundary Condition

Here we consider the constant production rate inner boundary condition, which is treated by adding the rate term directly to the grid fracture pressure control equation of wells.

8.1.2.5. Outer Boundary Condition

Closed outer boundary and constant pressure outer boundary are included, which are described in Equations 8.1.11 and 8.1.12.

$$\left.\begin{array}{l} \dfrac{\partial p_f}{\partial n}\bigg|_\Gamma = 0 \\[2mm] \dfrac{\partial p_m}{\partial n}\bigg|_\Gamma = 0 \end{array}\right\} \qquad (8.1.11)$$

$$\left.\begin{array}{l} p_f|_\Gamma = p_e \\ p_m|_\Gamma = p_e \end{array}\right\} \qquad (8.1.12)$$

8.1.3. The Streamline Mathematical Model of the Testing Period

By solving the filtration mathematical model of the production period for a multi-layer reservoir in dual-porosity media, pressure and saturation distribution of the fracture and matrix systems are obtained. Then, streamline tracing begins with the testing well based on the pressure distribution in the fracture system, and the streamline distribution of the testing well at the testing period is established. Next, convert the saturation parameter of the fracture grids into each streamline node, and take them as the initial condition of the streamline mathematical model for the testing period. Assuming that the cross-flow between the matrix and fracture systems is pseudo-steady flow, the streamline mathematical model of the testing period for a multi-layer reservoir in dual-porosity media along each streamline is established as follows.

For a multi-layer reservoir with N layers perforated, superscript n represents layer order, subscript j represents streamline order. Assuming that the streamlines from the testing well is N^n in nth layer, the flow control equation along each streamline is written as follows:

$$\frac{1}{l_j^n}\frac{\partial}{\partial l_j^n}\left(\alpha l_j^n \frac{\lambda_{tfj}^n}{\phi_{fj}^n}\frac{\partial p_{fj}^n}{\partial l_j^n}\right) = \alpha\omega c_{tf}\frac{\partial p_{fj}^n}{\partial t}$$

$$+ \alpha(1-\omega)c_{tf}\frac{\partial p_{mj}^n}{\partial t} \quad (n=1,2,\cdots,N; j=1,2,\cdots,N^n) \quad (8.1.13)$$

$$(1-\omega)\frac{\partial p_{mj}^n}{\partial t} = \beta(p_{fj}^n - p_{mj}^n)\,(j=1,2,\cdots,N) \quad (8.1.14)$$

Where λ_{tfj}^n is total mobility of jth streamline in nth layer fracture system; ω is elastic storativity ratio, $\omega = \frac{V_f\phi_{tf}C_f}{V_f\phi_{tf}C_f+V_m\phi_m C_{tm}}$ (dimensionless); β is inter-porosity cross-flow coefficient, $\beta = F_s\frac{K_m}{K_f}r_w^2$; V_f and V_m is the volume percentage of the fracture system and the matrix system, respectively (dimensionless).

Inner boundary condition: well bore storage effect and skin effect are considered. For well bore storage inner boundary, the production well and the injection well are included, which is described in Equations 8.1.15 and 8.1.16, respectively:

$$\sum_{n=1}^{N}\sum_{j=1}^{N^n}\frac{2\pi}{N^n}r_w h_1^n\left(\lambda_{tfj}^n\frac{\partial p_{fj}^n}{\partial l_j^n}\right)_{l_j^n=r_w}$$

$$= -q + C\frac{dp_{wf}}{dt} \quad (n=1,2,\cdots,N; j=1,2,\cdots,N^n) \quad (8.1.15)$$

$$\sum_{n=1}^{N}\sum_{j=1}^{N^n}\frac{2\pi}{N^n}r_w h_1^n\left(\lambda_{tfj}^n\frac{\partial p_{fj}^n}{\partial l_j^n}\right)_{l_j^n=r_w}$$

$$= q + C\frac{dp_{wf}}{dt} \quad (n=1,2,\cdots,N; j=1,2,\cdots,N^n) \quad (8.1.16)$$

Skin inner boundary condition (bottom-hole flow pressure simultaneous condition):

$$p_{wf} = p_{wj}^n - S^n\left(l_j^n\frac{\partial p_{fj}^n}{\partial l_{fj}^n}\right)_{l_j^n=r_w} \quad (n=1,2,\cdots,M; j=1,2,\cdots,N^n) \quad (8.1.17)$$

Outer boundary condition: this is the same as the filtration mathematical model of the production period.

Initial condition: this is obtained from the filtration mathematical model of the production period.

Saturation equation: Equations 8.1.18 and 8.1.19 describe the saturation equations for the fracture system and the matrix system of the testing period, respectively, which is the same as that of production period. The derivation and resolution method will be introduced in detail later.

$$\frac{\partial S_{wf}}{\partial t} + \frac{\partial f_{wf}}{\partial \tau} + \frac{\phi_m}{\phi_f}\frac{\partial S_{wm}}{\partial t} = 0 \tag{8.1.18}$$

$$T_w = \phi_m \frac{\partial S_{wm}}{\partial t} \tag{8.1.19}$$

8.2. SOLVING METHODS OF THE STREAMLINE NUMERICAL WELL TESTING INTERPRETATION MODEL FOR A MULTI-LAYER RESERVOIR IN DOUBLE-POROSITY MEDIA

8.2.1. Solving Method of the Filtration Mathematical Model in the Production Period

In the physical model of dual-porosity media, the flow in the matrix system is ignored, and block is seen as oil "source" for fracture system, then the diffusion term and source/sink term in Equation 8.1.6 is zero and $T_t = 0$, so Equation 8.1.3 converts as follows:

$$\nabla \cdot (K_f \lambda_{tf} \nabla p_f) = -q_{sf} \tag{8.2.1}$$

In dual-porosity media, there is no fluid flow or streamline between blocks, so only streamline tracing in fractures is enough. With the pressure equation of the fracture system 8.2.1, no cross-flow term appears in the equation. The cross-flow term will not affect the tracing path of the streamline, so the tracing method is the same with single-porosity media.

For the fracture system, the saturation equation is written as follows:

$$\phi_f \frac{\partial S_{wf}}{\partial t} + \vec{w}_{tf} \cdot \nabla f_{wf} + T_w = 0 \tag{8.2.2}$$

Operator $\vec{w}_t \cdot \nabla = \phi\frac{\partial}{\partial \tau}$ is applied into Equation 8.2.2 to realize the coordinate transformation for Equation 8.2.2, and the grid coordinate is converted to the form of streamline coordinate, namely Equation 8.2.3:

$$\frac{\partial S_{wf}}{\partial t} + \frac{\partial f_{wf}}{\partial \tau_f} + \frac{T_w}{\phi_f} = 0 \tag{8.2.3}$$

For the matrix system, the saturation equation is written as:

$$\phi_m \frac{\partial S_{wm}}{\partial t} + \vec{w}_{tm} \cdot \nabla f_{wm} - T_w = 0 \tag{8.2.4}$$

Convert grid coordinate into streamline coordinate:

$$\frac{\partial S_{wm}}{\partial t} + \frac{\partial f_{wm}}{\partial \tau_m} - \frac{T_w}{\phi_m} = 0 \tag{8.2.5}$$

Without considering the flow in the matrix system, the saturation equation in the matrix system is simplified to:

$$T_w = \phi_m \frac{\partial S_{wm}}{\partial t} \tag{8.2.6}$$

Equations 8.2.3 and 8.2.6 are saturation propagation equations of the fracture and the matrix systems, respectively.

Division and difference methods are used to solve the saturation equation of the fracture system. At first, the water saturation term in Equation 8.2.3 is divided into two terms, and Equation 8.2.7 is obtained; then, substitute Equation 8.2.7 into Equation 8.2.3 and get Equations 8.2.8 and 8.2.9; finally, discrete Equations 8.2.8 and 8.2.9 to get the different Equations 8.2.10 and 8.2.11, which are easy to solve.

$$\frac{\partial S_{wf}}{\partial t} = \frac{\partial S_{wf}}{\partial t_1} + \frac{\partial S_{wf}}{\partial t_2} \tag{8.2.7}$$

$$\frac{\partial S_{wf}}{\partial t_1} + \frac{\partial f_{wf}}{\partial \tau_f} = 0 \tag{8.2.8}$$

$$\frac{\partial S_{wf}}{\partial t_2} + \frac{T_w}{\phi_f} = 0 \tag{8.2.9}$$

$$S_{wf,i}^{n+1} - S_{sf,i}^n = -\frac{1}{t_{1,n+1} - t_{1,n}} \cdot \frac{f_{wf,i+\frac{1}{2}}^n - f_{wf,i-\frac{1}{2}}^n}{\Delta \tau_f} \tag{8.2.10}$$

$$S_{wf,i}^{n+1} - S_{sf,i}^n = -\frac{1}{t_{2,n+1} - t_{2,n}} \cdot \left[\left(\frac{F_s K_m}{\phi_f} \right)_i \left(\frac{\lambda_w \lambda_o}{\lambda_o + \lambda_w} \right)_i^n (p_m - p_f)_i^n \right] \tag{8.2.11}$$

The difference method is directly used to solve the saturation equation of the matrix system (Equation 8.2.6), and the difference equation is written as follows:

$$S_{wm,i}^{n+1} - S_{sm,i}^n = -\frac{1}{t_{n+1} - t_n} \cdot \left[\left(\frac{F_s K_m}{\phi_f} \right)_i \left(\frac{\lambda_w \lambda_o}{\lambda_o + \lambda_w} \right)_i^n (p_m - p_f)_i^n \right] \tag{8.2.12}$$

8.2.2. Solving Method of the Streamline Mathematical Model in the Testing Period

The streamline mathematical model of the testing period for a multi-layer reservoir in dual-porosity media is similar to that in single-porosity media, which is solved by combing all the streamlines in all layers of the testing wells together. For a testing well with M layers perforated, if the streamline in each layer is N, and there are N_J nodes on each streamline, the number of total equations in all discrete streamline models is $2 \times M \times N \times N_J$

8.3. PRESSURE RESPONSE CHARACTERISTICS OF THE STREAMLINE NUMERICAL WELL TESTING INTERPRETATION MODEL FOR A MULTI-LAYER RESERVOIR IN DUAL-POROSITY MEDIA

8.3.1. Effect of Inter-Layer Permeability Ratio on Pressure Response

In order to study the influencing characteristics of inter-layer permeability ratio (the ratio of maximum permeability to minimum permeability) on the streamline numerical well testing interpretation model for a multi-layer reservoir in dual-porosity media, an inverted five-spot typical model of three layers is established, and the main parameters are described below.

The porosity is 0.2, oil formation thickness is 10 m, and elastic storativity for the fracture system is 0.024, cross-flow coefficient between matrix and fracture is 6.4×10^{-8}, physical parameters and phase permeability of the oil phase are the same as those of the water phase, and the multi-phase flow model is degenerated into a single-phase flow model. Fluid production of oil well P1~P3 is $80 \, m^3 / d$, fluid production of oil well P4 is $60 \, m^3 / d$, fluid injection of water well I is $100 \, m^3 / d$. All the oil and water wells in the three layers are perforated.

At first, the effect of permeability ratio on streamline numerical well testing interpretation model for multi-layer reservoir in dual porosity media is studied without interlayer crossflow.

Figure 8.2 shows the pressure response curve with the same permeability average value and permeability ratio 1, 9, 25, 100, respectively. As seen in this figure, the pressure derivation curve appears "downward concave" during the radial flow period; and the "downward concave" becomes more serious with the increase in permeability ratio. The "downward concave" phenomenon of pressure derivation is caused by cross-flow between matrix and fracture, and the inter-layer heterogeneity aggravates this phenomenon.

Then, effect of permeability ratio on the streamline numerical well testing interpretation model for a multi-layer reservoir in dual-porosity media is studied with inter-layer cross-flow.

Figures 8.3 to Fig. 8.5 show streamline distribution with the same average permeability and permeability ratio 9, 25 and 100, respectively. Figure 8.6 shows the pressure response curve contrast.

As seen in Fig. 8.6, with inter-layer cross-flow, the pressure derivation curve appears "downward concave" not only in the middle-radial flow period, but also in the later period. As analyzed above, the "downward concave" of the pressure derivation curve in the middle radial period is caused by cross-flow between matrix and fracture, and the "downward concave" of pressure derivation in the later period is caused by inter-layer cross-flow in a multi-layer reservoir, the larger the permeability ratio is, the more serious the cross-flow will be.

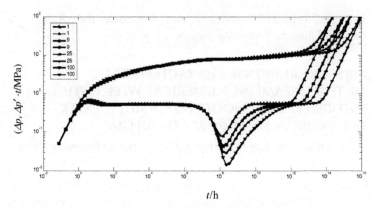

FIGURE 8.2   Pressure response contrast of different inter-layer permeability ratio without inter-layer cross-flow.

8.3.2. Effect of Elastic Storativity on Pressure Response

The pressure response influencing characteristic of elastic storativity on the streamline numerical well testing interpretation model for a multi-layer reservoir in dual-porosity media is studied based on the typical model above.

Figure 8.7 shows pressure response curve without inter-layer cross-flow and the elastic storativity is 0.02, 0.04, 0.06 and 0.09, respectively. It can be seen that, the "downward concave" of pressure derivation becomes weaker with the increase in elastic storativity.

8.3.3. Effect of Cross-Flow Factor on Pressure Response

The influencing characteristic of cross-flow coefficient between fracture and matrix in the streamline numerical well testing interpretation model for a

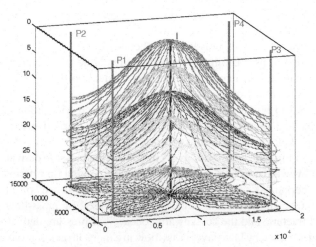

FIGURE 8.3   Streamline distribution with inter-layer permeability 9.

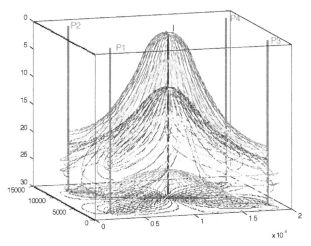

FIGURE 8.4   Streamline distribution with inter-layer permeability 25.

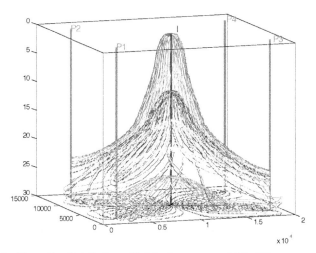

FIGURE 8.5   Streamline distribution with inter-layer permeability 100.

multi-layer reservoir in dual-porosity media is still studied based on the typical model above.

Figure 8.8 shows the pressure response curve without inter-layer cross-flow and the cross-flow coefficient is 1.6×10^{-8}, 6.4×10^{-8} and 2.56×10^{-7}, respectively. It can be seen that the start time of the "downward concave" in pressure derivation starts earlier with the increase in cross-flow coefficient.

8.4. CHAPTER SUMMARY

In this chapter, the streamline numerical well testing interpretation model for a multi-layer reservoir in dual-porosity media is studied based on the streamline

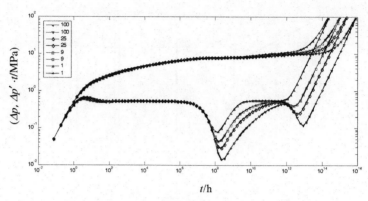

FIGURE 8.6   Pressure response contrast of different permeability ratio with inter-layer cross-flow.

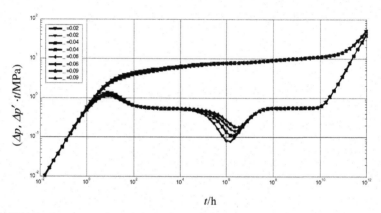

FIGURE 8.7   Pressure response contrast with different elastic storativity.

numerical well testing interpretation model for a multi-layer reservoir in single-porosity media. The main contents are shown as follows:

1. The streamline numerical well testing interpretation model for a multi-layer reservoir in dual-porosity media includes the filtration mathematical model of the production period and the streamline mathematical model of the testing period: the filtration mathematical model of the production period adopts the simplified black oil model for a multi-layer reservoir in dual-porosity media, which could consider the cross-flow between matrix and fracture and inter-layer cross-flow; the mathematical model of the testing period is the streamline model, which is established by combining all the mathematical equations of each streamline from all layers in the fracture system. The streamline mathematical model of the testing period is established with the pressure and saturation distribution as fundamental and initial conditions which are solved during the production period.

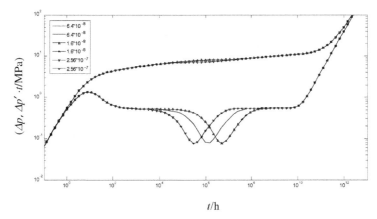

FIGURE 8.8 Pressure response contrast with different cross-flow coefficient.

2. The solution to the filtration mathematical model of the production period adopts the combination of the IMPES and the streamline methods. At first, the IMPES method is used to solve the pressure distribution of the matrix and fracture systems in the research region; then, trace the streamline and establish the saturation equation along the streamline based on the pressure distribution of the fracture system; finally, solve the saturation equation along the streamline and obtain the saturation distribution of the testing well during the testing period.

3. The solution to the streamline mathematical model of the testing period is similar to that of single porosity media, namely, solve together with all the streamlines in all layers of the testing well.

4. Through the pressure drawdown simulation of the testing well in a typical model of a multi-layer reservoir for dual-porosity media, the effect of permeability ratio, elastic storativity and cross-flow coefficient on pressure response of the testing well are studied:

   - when inter-layer cross-flow does not exist, the pressure derivation curve appears "downward concave" and this is caused by cross-flow between the matrix and fracture systems;
   - when inter-layer cross-flow exists, the pressure derivation curve appears "downward concave" twice under certain conditions; the first is caused by cross-flow between the matrix and fracture systems, the second is caused by inter-layer cross-flow;
   - increase in permeability ratio aggravates cross-flow between the matrix and the fracture systems and inter-layer cross-flow;
   - the larger the elastic storativity of the fracture system is, the weaker the downward concave of pressure derivation will be, which is caused by cross-flow between the matrix and fracture systems.
   - the larger the storativity coefficient is, the earlier the start time of the downward concave of pressure derivation will be, which is caused by cross-flow between the matrix and fracture systems.

Streamline Numerical Well Testing Interpretation Model of a Horizontal Well

9.1. ESTABLISHMENT OF THE STREAMLINE NUMERICAL WELL TESTING INTERPRETATION MODEL OF A HORIZONTAL WELL

The streamline numerical well testing interpretation model of a horizontal well is made up of the filtration mathematical model of the production period and the streamline mathematical model of the testing period. The filtration mathematical model of the production period takes the simplified black oil model, which is used to calculate the pressure distribution at the testing time and to obtain the streamlines of the testing region. The streamline mathematical model of the testing period is established along each streamline from the test well, which takes the pressure and saturation obtained in the production period as the initial condition. The streamline mathematical model is composed of the equations of each streamline at the testing period.

9.1.1. Filtration Mathematical Model of the Production Period
9.1.1.1. Mathematical Model

The filtration mathematical model of a horizontal well in the production period is based on the following basic assumptions:

1. Rocks and fluids are incompressible.
2. The reservoir is heterogeneous and permeability is anisotropic.
3. Oil and water phases exist in the reservoir and obey the Darcy law.
4. The impact of gravity and capillary force are ignored.
5. The horizontal well produces at the condition of uniform flow rate or constant flow pressure, the well bore flow considers the skin effect and well bore storage effect and ignores the friction loss along the horizontal well bore.
6. The formation pressure is non-uniformly distributed before the testing.

Streamline Numerical Well Test Interpretation. DOI: 10.1016/B978-0-12-386027-9.00009-5
Copyright © 2011 by Elsevier Ltd

Oil-phase flow equation

$$\nabla \cdot \left(\frac{\rho_o K K_{ro}}{\mu_o} \nabla p\right) + q_o = \frac{\partial(\phi \rho_o S_o)}{\partial t} \tag{9.1.1}$$

Water-phase flow equation

$$\nabla \cdot \left(\frac{\rho_w K K_{rw}}{\mu_w} \nabla p\right) + q_w = \frac{\partial(\phi \rho_w S_w)}{\partial t} \tag{9.1.2}$$

$$S_o + S_w = 1 \tag{9.1.3}$$

Initial condition

$$\left.p\right|_{t=0} = p_{il} \tag{9.1.4}$$
$$\left.S_w\right|_{t=0} = S_{wil} \tag{9.1.5}$$

Outer boundary condition

$$\left.\frac{\partial p}{\partial n}\right|_\Gamma = 0 \tag{9.1.6}$$

$$\left.p\right|_\Gamma = p_e \tag{9.1.7}$$

9.1.1.2. Establishment of Horizontal Well Inner Boundary Conditions
9.1.1.2.1. The Well Bore Inflow Performance Relationship Description of a Horizontal Well

There are mainly two methods to describe the well bore inflow performance relationship of a horizontal well: the radial flow equation of the vertical well and the elliptic flow equation.

(i) The Peaceman model is most commonly used to describe the inflow performance of the horizontal well when using the radial flow equation of the vertical well, which simulates the horizontal well with vertical well by rotating the coordinate axis. Assume the horizontal well is parallel to the y-axis, exchange the y-axis and z-axis and then the output of the horizontal well is

$$q = \sum_{m=1}^{M} \frac{-2\pi(k_x/k_z)^{1/2}(k_x k_z)^{1/2}\Delta y(p_m - p_{wf})}{\mu[\ln(r_e/r_w) + S]} \tag{9.1.8}$$

where:

p_m is the grid pressure (MPa);
p_w is the bottom-hole pressure (MPa);
S is the skin factor, dimensionless;
k_x and k_z are the absolute permeabilities in the x and z directions, respectively (μm^2);
Δx and Δz are the grid spacings in x and z directions, respectively (cm);
M is the number of grids crossed by the horizontal well;
r_e is the equivalent supply radius of grid block (cm), which is determined by the following formula.

$$r_e = 0.28 \frac{k_x/k_z^{1/2}\Delta Z^2 + (k_z/k_x)^{1/2}\Delta x^2}{(k_x/k_z)^{1/4} + (k_z/k_x)^{1/4}} \tag{9.1.9}$$

(ii) When the well bore inflow performance of the horizontal well is described by the elliptic flow equation, the output of the horizontal well is

$$q = \sum_{m=1}^{M} \frac{-2\pi(k_x/k_z)^{1/2}(k_xk_z)^{1/2}h(p_m - p_{wf})}{\mu\left\{\ln\left(\frac{a+[a^2-(L/2)^2]^{1/2}}{L/2}\right) + \frac{(k_xk_z)^{1/2}h}{L}\ln\left(\frac{(k_xk_z)^{1/2}(h/2)^2+\delta^2}{hr_{we}/2}\right)\right\}} \tag{9.1.10}$$

where:

h is the effective thickness of the reservoir (cm);
L is the horizontal section length of the horizontal well (cm);
δ is the vertical distance between the horizontal well and reservoir center (cm);
a is the primary semi-axis length of the elliptical drainage plane of the horizontal well (cm); r_{we} is the equivalent radius (cm).

a is determined by

$$a = (L/2)\left[0.5 + +\sqrt{0.25 + (2r_e/L)^4}\right]^{1/2} \tag{9.1.11}$$

r_{we} is determined by

$$r_{we} = \frac{r_eL/2}{a\left[1 + \sqrt{1 - (L/2a)^2}\right](h/2r_w)^{(k_x/k_z)h/L}} \tag{9.1.12}$$

9.1.1.2.2. Inner Boundary Conditions
The inner boundary conditions of the horizontal well include constant pressure and constant rate in two cases.

(i) Inner boundary condition of constant pressure: when the horizontal well produces at constant flow pressure, according to the inflow performance relationship Equation 9.1.8 or 9.1.10 above, the output of each grid crossed by the horizontal well can be directly calculated by

$$q_m \frac{-2\pi(k_x/k_z)^{1/2}(k_xk_z)^{1/2}\Delta y(p_m - p_{wf})}{\mu[\ln(r_e/r_w) + S]} \tag{9.1.13}$$

where, q_m is the production item of the grid crossed by the horizontal well, which can be brought into the flow equation.

(ii) Inner boundary condition of constant production rate: when the horizontal well produces at constant production rate, the flow pressure $p_{wf_{ref}}$ at a certain reference depth (usually $0.65 \sim 0.71$ l) of the horizontal well is calculated by Equation 9.1.8 or 9.1.10. Distribute the production of each grid by Equation 9.1.13, and then the production item of each grid can be bought into the flow equation.

$$q_m = q \cdot \frac{\frac{-2\pi(k_x/k_z)^{1/2}(k_x k_z)^{1/2}\Delta y \left(p_m - p_{wf_{ref}}\right)}{\mu[\ln(r_e/r_w)+S]}}{\sum_{m=1}^{M} \frac{-2\pi(k_x/k_z)^{1/2}(k_x k_z)^{1/2}\Delta y \left(p_m - p_{wf_{ref}}\right)}{\mu[\ln(r_e/r_w)+S]}} \tag{9.1.14}$$

The calculation method of the production item above is used for the horizontal production well. The sign of q should be opposite when the horizontal well is injection well.

9.1.2. Streamline Mathematical Model of the Testing Period

Filtration control equation: assume M is the number of grids crossed by the test well; N_m is the sum of the streamlines from each grid. Then the filtration control equation along each streamline is

$$\frac{1}{l_j^m}\frac{\partial}{\partial l_j^m}\left(l_j^m \frac{\lambda_t}{\phi}\frac{\partial p_j^m}{\partial l_j^m}\right) = c_t \frac{\partial p_j^m}{\partial t} \tag{9.1.15}$$

where:

l is the curve distance along the streamline with the test well location as the origin point (cm);
p is the pressure along streamline (10^{-1} MPa);
c_t is the total compressibility factor of reservoir rock and fluids (10MPa^{-1});
λ_t is the total mobility of oil and water phase [$\mu m^2/(mPa\bullet s)$]; $j = 1,2,\ldots,$
N_m; $m = 1,2,\ldots,M$.

Initial condition

$$p(l_j, t)|_{t=0} = p_{i2} \tag{9.1.16}$$
$$S_w(l_j, t)|_{t=0} = S_{wi2} \tag{9.1.17}$$

where:

p_{i2} is the reservoir pressure at the testing time (10^{-1} MPa);
S_{wi} is the reservoir water saturation at the testing time (decimal);
p_{i2} and S_{wi2} are obtained by solving the well-testing interpretation model of production period.

Inner boundary condition: the inner boundary condition of the model consists of two conditions, namely the infinite conductivity condition and uniform flow condition. In this paper we use the horizontal production well as the example to elaborate upon. It is similar to the case when the test well is an injection well, which just has a negative sign in front of the production item in the equation. The details will not be given here.

(i) Inner boundary condition of infinite conductivity: when the horizontal well has infinite conductivity, the flow pressure of each horizontal section is equal but the output is unequal. The inner boundary condition at this time is

$$\sum_{m=1}^{M}\sum_{j=1}^{N_m}\frac{2\pi}{N_m}r_w L^m \lambda_t \left(\frac{\partial p_j^m}{\partial l_j^m}\right)_{l_j=r_w} = -q + C\frac{dp_{wf}}{dt} \tag{9.1.18}$$

The flow pressure of the horizontal well is

$$p_{wf} = p_{wj}^m - Sr_w\left(\frac{\partial p_j^m}{\partial l_j^m}\right)_{l_j=r_w} \tag{9.1.19}$$

The production of each horizontal section is

$$q^m = \sum_{j=1}^{N_m}\frac{2\pi}{N_m}r_w L^m \lambda_t \left(\frac{\partial p_j^m}{\partial l_j^m}\right)_{l_j=r_w} \tag{9.1.20}$$

$$q = \sum_{m=1}^{M}q^m = \sum_{m=1}^{M}\sum_{j=1}^{N_m}\frac{2\pi}{N_m}r_w L^m \lambda_t \left(\frac{\partial p_j^m}{\partial l_j^m}\right)_{l_j=r_w} \tag{9.1.21}$$

where:
p_w is the borehole pressure of the testing well, which is also the first node pressure of each streamline (10^{-1} MPa);
q is the constant production (injection) rate of the test well at well open test, or the constant production (injection) rate before the shut-in test (cm^3/s);
C is the well bore storage coefficient [$cm^3/(10^{-1}$ MPa)];
r_w is the radius of the testing well (cm);
L^m is the horizontal well length in the mth grid (cm).

(ii) Inner boundary condition of uniform flow: when the horizontal well produces at uniform flow conditions, the production of each horizontal section is equal but the flow pressure is unequal. The inner boundary condition at this time is

$$\sum_{m=1}^{M}\sum_{j=1}^{N_m}\frac{2\pi}{N_m}r_w L^m \lambda_t \left(\frac{\partial p_j^m}{\partial l_j^m}\right)_{l_j=r_w} = -q + C\frac{dp_{wf}^m}{dt} \tag{9.1.22}$$

Flow pressure of each horizontal section is

$$p_{wf}^m = \left(p_{wf}^m - Sr_w\frac{\partial p_j^m}{\partial l_j^m}\right)_{l_j=r_w} \tag{9.1.23}$$

The production of each horizontal section is

$$q^m = q/M \tag{9.1.24}$$

Outer boundary condition: the outer boundary of the model (the end of streamline) can be divided into the oil (water) well outer boundary and reservoir outer boundary according to the end location of streamline. The oil (water) well outer boundary refers to the case when streamline starts from the test well and stops at other oil (water) wells. The outer boundary condition at this time is the working condition of the oil (water) wells at the end of streamline. The reservoir outer boundary refers to the case when streamline ends at the reservoir boundary, which includes the constant pressure boundary and closed boundary. The reservoir outer boundary condition is the same with the conventional well testing interpretation model.

9.2. SOLVING METHODS OF THE STREAMLINE NUMERICAL WELL TESTING INTERPRETATION MODEL OF A HORIZONTAL WELL AND VERIFICATION

9.2.1. Solving Method of the Filtration Mathematical Model of the Production Period

The filtration mathematical model of the production period is solved by combining the IMPES method and the streamline method. First of all, the differential divergence method is applied to the model, and the model is solved by the IMPES method. Then the pressure distribution of the grid system at the shut-in test time is obtained. On this basis, the streamline tracing method proposed by Pollock is used to trace the streamline distribution in the test area at the shut-in time. With the travel-time method, grid parameters at shut-in time which are solved in the production period are transformed to each node of streamlines. The mobility value of each node is calculated and the saturation equation along the streamline is established. Finally, the saturation equations of all streamlines are solved and the saturation distribution in the study area is obtained.

9.2.2. Solving Method of the Streamline Mathematical Model in the Testing Period

Diverge Equations 9.1.15 to 9.1.24 with the differential method, collect all the differential equations of all streamlines and derive the following tridiagonal equation.

$$b_{j,i} p_{j,i-1}^{n+1} + c_{j,i} p_{j,i}^{n+1} + d_{j,i} p_{j,i+1}^{n+1} = g_{j,i} \tag{9.2.1}$$

where $b_{j,i}; d_{j,i}; c_{j,i}; g_{j,i}$ are divergence coefficients; $j = 1,2,\ldots,N_m$; $i = 1,2,\ldots,N_j$ (N_j is the node number of the jth streamline).

The theoretical pressure response after the testing well shut-in can be obtained by solving the linear Equation 9.2.1 with the iterative method or chasing algorithm. The well test interpretation parameters of the testing area could also be obtained by automatic matching between the theoretical pressure response and measured pressure data with dual-population genetic algorithm.

9.2.3. Verification of the Streamline Numerical Well Testing Method of a Horizontal Well

The typical horizontal–vertical hybrid inverted five-spot well pattern in a single-layer reservoir is established. The pressure response characteristics of the streamline numerical well testing interpretation model of a horizontal well are studied.

Basis parameters of reservoir and wells: the length of reservoir is 1640 m, and the width is 1640 m; reservoir effective thickness is 60 m; the length of the horizontal section of the horizontal well is 700 m; reservoir porosity is 0.25; horizontal permeability and vertical permeability are $0.1\,\mu m^2$ and $0.08\,\mu m^2$, respectively; the liquid production rate of production wells is $40\,m^3/d$; the injection rate of injection well is $160\,m^3/d$.

Shut in the horizontal well and simulate the pressure drawdown, and reduce the two-phase flow into single-phase flow. The pressure and streamline distribution at the testing time are shown in Figs 9.1 and 9.2. Figure 9.3 presents the streamline and pressure distribution in a different profile and Fig. 9.4 shows the theoretical pressure response of the testing well.

It can be seen from Fig. 9.4 that the flow phases of the horizontal well can be divided as:

1. The well bore storage phase: the pressure and derivative curve show a unity slope due to the well bore storage effect.
2. The early radial flow phase: it is radial flow in the vertical plane, the streamlines are radially distributed with the well bore as center in the near-well-bore vertical plane, and the pressure derivative curve shows a horizontal line in the log–log graph.
3. The early linear flow phase: the streamlines arrive at the upper and lower boundary and distribute linearly in the formation (Fig. 9.3c and d), the pressure derivative shows a straight line with 0.5 slope in the log–log graph.

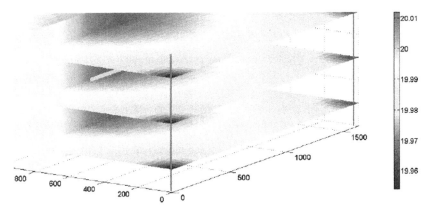

FIGURE 9.1 Pressure profile at the testing time.

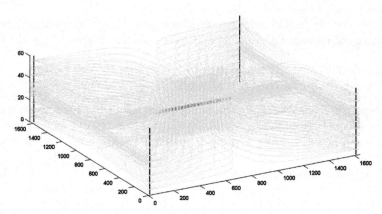

FIGURE 9.2   Streamline distribution at the testing time.

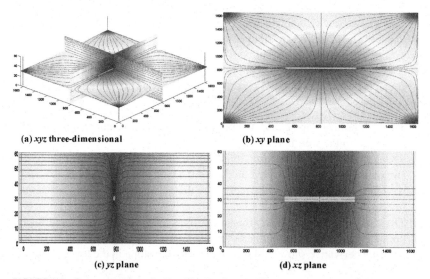

(a) *xyz* three-dimensional   (b) *xy* plane

(c) *yz* plane   (d) *xz* plane

FIGURE 9.3   Pressure and streamline distribution profile.

4. The late radial flow phase: streamlines assemble at the horizontal well from all directions and form the radial flow in the horizontal plane, and the pressure derivative shows a horizontal line in the log–log graph.
5. Pseudo-steady state flow phase: the streamlines reach the outer boundary (Fig. 9.3b–d) and the pressure derivative curve shows an upward tendency.

So the results calculated from the streamline numerical well testing model of the horizontal well conform to the flow pattern of horizontal well, which verifies the validity of the streamline numerical well testing interpretation model of the horizontal well.

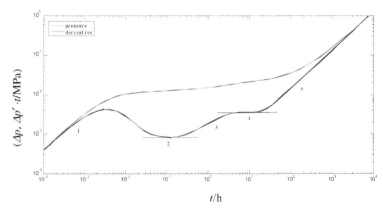

FIGURE 9.4   Pressure response characteristic curves of the testing well.

9.3. PRESSURE RESPONSE CHARACTERISTICS OF THE STREAMLINE NUMERICAL WELL TESTING INTERPRETATION MODEL OF A HORIZONTAL WELL

9.3.1. Influencing Characteristics of Skin Factor on Pressure Response

In order to study the influencing characteristics of well bore skin factor on pressure response of horizontal well streamline numerical well testing, the model is degenerated into single-phase flow model. The basic reservoir parameters are shown in Table 9.1.

By solving the typical well testing interpretation model of the horizontal well above, we can obtain the pressure profile (Fig. 9.5), streamline distribution (Fig. 9.6), streamline profile (Fig. 9.7) at the testing time and the pressure response curves with different skin factors (Fig. 9.8), respectively.

TABLE 9.1 Basic Parameters

Reservoir parameters	Parameter value	Reservoir parameters	Parameter value
Length* width* thickness of reservoir (m)	4100*4100*240	Volume factor (m^3/m^3)	1.18
Porosity	0.3	Total compressibility (MPa^{-1})	0.003
Oil viscosity (mPa•s)	4	Daily oil production (m^3/d)	60
Horizontal permeability ($10^{-3}\,\mu m^2$)	50	Radius of horizontal well (m)	0.08
Vertical permeability ($10^{-3}\,\mu m^2$)	50	Daily water injection (m^3/d)	240

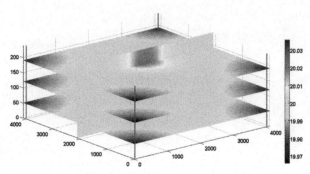

FIGURE 9.5   Pressure profile at the testing time.

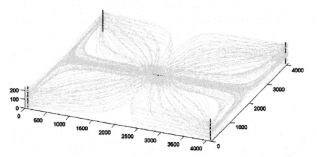

FIGURE 9.6   Streamline distribution at the testing time.

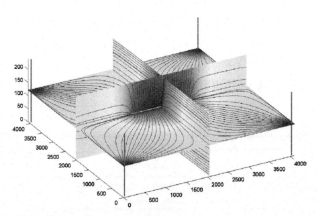

FIGURE 9.7   Streamline profile at the testing time.

As seen from the above pressure and derivative curve, with the increase of skin factor, the pressure and derivative curves move upward, the bossy peak of the pressure derivative at the beginning stage is bigger and the duration of the radial flow phase is shorter. This is because the larger the skin factor, the more serious the borehole damage, which causes a bigger additional pressure drop. Additionally, with the increase in skin factor, the pollution area becomes larger and the time of fluids flowing through the damage zone becomes greater. So the time of

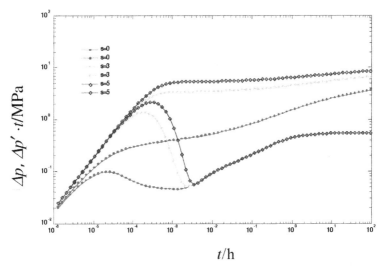

FIGURE 9.8 Pressure response curves with different skin factors.

the pressure response entering the next flow stage is delayed, which means that the time of peak reaching the highest point in the graph is also delayed.

9.3.2. Influencing Characteristics of Well Bore Storage Factor on Pressure Response

Take the model used above as an example, simulate the pressure drawdown test of the horizontal well and obtain the pressure response curves with different well bore storage coefficients (Fig. 9.9). As seen from Fig. 9.9, with the increase of well bore storage coefficient, the pressure curves move to the right entirely, which delays the time of the pressure response entering the next flow stage. This is because the larger the well bore storage coefficient, the longer the well bore after flow lasts, so pressure change delays.

9.3.3. Influencing Characteristics of Reservoir Relative Thickness on Pressure Response

The influencing characteristics of the reservoir relative thickness (the ratio of reservoir thickness to the horizontal section length of horizontal well) on pressure response are studied. The model is degenerated into single-phase flow where the skin factor is 0 and the well bore storage coefficient is $0.001 \, \text{m}^3 /$ MPa. The basic reservoir parameters are shown in Table 9.2.

By solving the typical well testing interpretation model of the horizontal well above, we can obtain the pressure profile (Fig. 9.10), the streamline distribution (Fig. 9.11), the streamline profile (Fig. 9.12) at the testing time and the pressure response curves with different reservoir relative thickness (Fig. 9.13), respectively.

As seen in Fig. 9.13, with the increase in the relative thickness, the duration of the horizontal well early radial flow increases, and the emergence time of the early

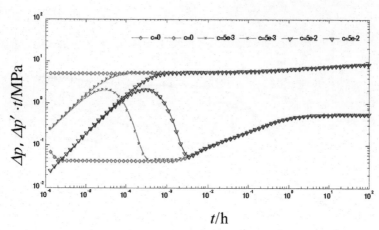

FIGURE 9.9   Pressure response curves with different well bore storage factors.

linear flow is delayed. This is because the horizontal well early radial flow happens in the plane vertical to the hole axis, with the increase of reservoir relative thickness, the area of the radial flow becomes larger, so the time lengthens. At the same time, the time taken for pressure response to reach the upper and lower boundary also becomes longer and so the emergence time of linear flow is delayed.

9.3.4. Influencing Characteristics of Horizontal Well Relative Position on Pressure Response

Assume that the distance from the horizontal well to the bottom of the reservoir is z_w and the effective thickness of the reservoir is h. The influencing characteristics

TABLE 9.2 Basic Parameters

Reservoir parameters	Parameter value	Reservoir parameters	Parameter value
Length* width* thickness of reservoir (m)	4100*4100*150	Volume factor (m^3/m^3)	1.18
Porosity	0.2	Total compressibility (MPa^{-1})	0.001
Oil viscosity (mpa·s)	1	Daily oil production (m^3/d)	50
Horizontal permeability $(10^{-3}\,\mu m^2)$	10	Radius of horizontal well (m)	0.08
Vertical permeability $(10^{-3}\,\mu m^2)$	10	Daily water injection (m^3/d)	200

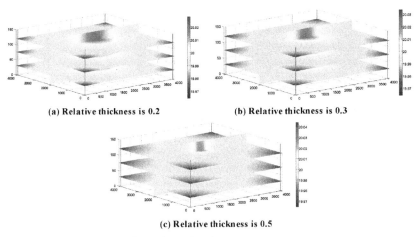

(a) Relative thickness is 0.2   (b) Relative thickness is 0.3

(c) Relative thickness is 0.5

FIGURE 9.10   Pressure profile at the testing time.

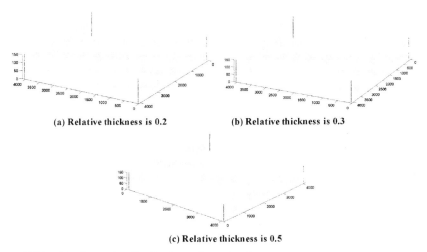

(a) Relative thickness is 0.2   (b) Relative thickness is 0.3

(c) Relative thickness is 0.5

FIGURE 9.11   Streamline distribution at the testing time.

of the relative position (z_w/h) on horizontal well streamline numerical well testing pressure response are studied. The model is degenerated into single-phase flow where the skin factor is 0 and the well bore storage coefficient is 0.001 m³ / MPa. The basic reservoir parameters are shown in Table 9.2. By solving this typical well testing interpretation model of the horizontal well, we can obtain the pressure profile (Fig. 9.14), the streamline distribution (Fig. 9.15) at the testing time and the pressure response curves with different horizontal well relative positions (Fig. 9.16), respectively.

As seen in Fig. 9.16, with the increase in the distance from the horizontal well to the bottom, the pressure and derivative curves at the early radial flow stage become longer and the duration of the early radial flow increases. When

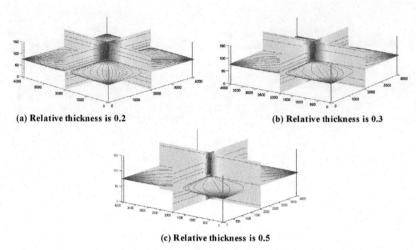

(a) Relative thickness is 0.2   (b) Relative thickness is 0.3

(c) Relative thickness is 0.5

FIGURE 9.12   Streamline profile at the testing time.

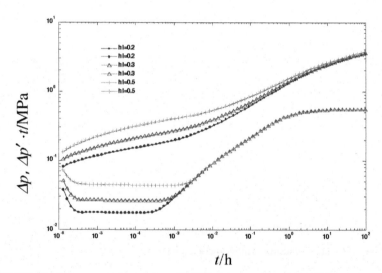

FIGURE 9.13   Pressure response curves with different reservoir relative thickness.

the relative position value is 0.1, there is a horizontal segment with 0 slope on the pressure response curve before the early linear flow occurs. It is because the horizontal well is close to the reservoir bottom and the hemiradial flow appears.

9.3.5. Influencing Characteristics of Permeability Ratio on Pressure Response

The influencing characteristics of the permeability ratio (the ratio of vertical permeability to horizontal permeability) on the pressure response of horizontal

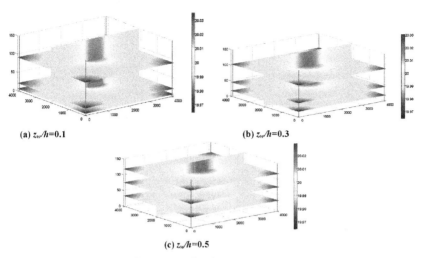

(a) $z_w/h=0.1$ (b) $z_w/h=0.3$

(c) $z_w/h=0.5$

FIGURE 9.14 **Pressure profile at the testing time.**

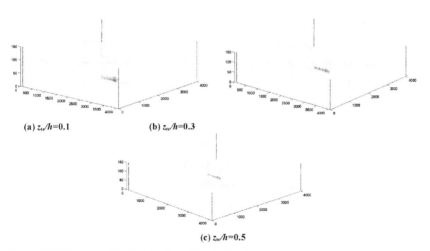

(a) $z_w/h=0.1$ (b) $z_w/h=0.3$

(c) $z_w/h=0.5$

FIGURE 9.15 **Streamline distribution at the testing time.**

well streamline numerical well testing are studied. The model is degenerated into single-phase flow where the skin factor is 0 and the well bore storage coefficient is $0.001 \, \text{m}^3 / \text{MPa}$. The basic reservoir parameters are shown in Table 9.2. By solving this typical well testing interpretation model of the horizontal well, we can obtain the pressure profile (Fig. 9.17), the streamline distribution (Fig. 9.18), the streamline profile (Fig. 9.19) at the testing time and the pressure response curves with different permeability ratios (Fig. 9.20), respectively.

As seen in Fig. 9.20, with the increase in permeability ratio, the pressure and derivative curves move downwards, the duration of early radial flow decreases and the early linear flow occurs earlier. This is because the fluid mobility

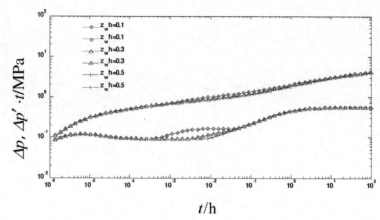

FIGURE 9.16   **Pressure response curves with different horizontal well relative positions.**

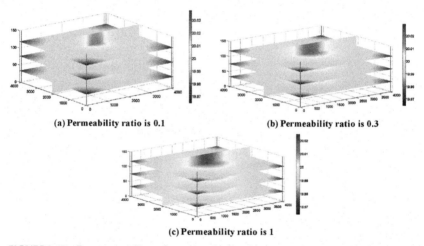

(a) Permeability ratio is 0.1

(b) Permeability ratio is 0.3

(c) Permeability ratio is 1

FIGURE 9.17   **Pressure profile at the testing time.**

increases with the increase in vertical permeability, and so the time for the pressure to reach the upper and lower boundary shortens.

9.3.6. Influencing Characteristics of Oil/Water Viscosity Ratio on Pressure Response

Take the inverted five-spot homogeneous reservoir as an example, the water injection rate of the horizontal well is $80\,m^3/d$ and the liquid production rate of the four production wells is $20\,m^3/d$. Simulate the pressure drawdown test of the horizontal well with different oil/water viscosity ratios under the same production conditions. The pressure response curves are shown in Fig. 9.21.

As seen in Fig. 9.21, with the increase in oil viscosity (the viscosity of water is fixed), the pressure response curves rise. This is because the increase in oil

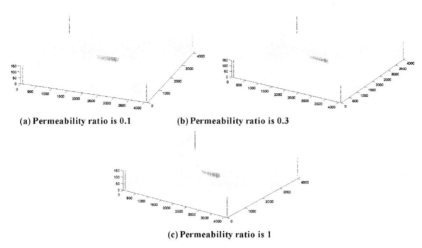

(a) Permeability ratio is 0.1   (b) Permeability ratio is 0.3

(c) Permeability ratio is 1

FIGURE 9.18   Streamline distribution at the testing time.

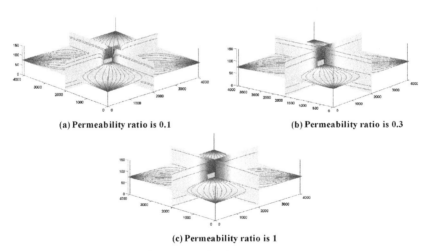

(a) Permeability ratio is 0.1   (b) Permeability ratio is 0.3

(c) Permeability ratio is 1

FIGURE 9.19   Streamline profile at the testing time.

viscosity makes the total oil–water two-phase mobility decrease, which results in the increase in flow resistance.

9.3.7. Pressure Response Characteristics in Different Well Patterns

Finally, the pressure response characteristics of horizontal well streamline numerical well testing in different well patterns (inverted five-spot pattern, inverted seven-spot pattern, and inverted nine-spot pattern) are studied. The basic reservoir parameters are shown in Tables 9.3 to 9.5.

By solving the typical well testing interpretation models of the horizontal well above, we can obtain the pressure profile (Fig. 9.22), the streamline

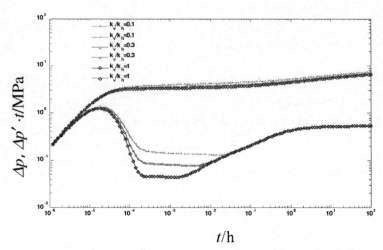

FIGURE 9.20   Pressure response curves with different permeability ratios.

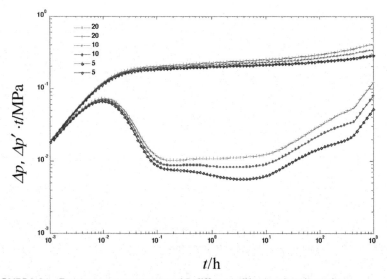

FIGURE 9.21   Pressure response curves with different oil/water viscosity ratios.

distribution (Fig. 9.23), the streamline profile (Fig. 9.24) at the testing time and the pressure response curves (Fig. 9.25), respectively.

As seen in Fig. 9.24, when pressure response spreads in the oil–water two-phase region, the pressure derivative curve shows upwarp tendency. It is because the fluid mobility decreases when the pressure transmits from water phase region into oil–water two-phase region. So the highest point of the pressure derivative upwarp is the water/oil front.

TABLE 9.3 Basic Parameters for Inverted Five-Spot Pattern

Reservoir parameters	Parameter value	Reservoir parameters	Parameter value
Length* width* thickness of reservoir (m)	610*610*30	Volume factor (m^3/m^3)	1.18
Porosity	0.2	Total compressibility (MPa^{-1})	0.003
Oil viscosity (mPa•s)	5	Daily oil production (m^3/d)	50
Horizontal permeability $(10^{-3}\,\mu m^2)$	100	Radius of horizontal well (m)	0.08
Vertical permeability $(10^{-3}\,\mu m^2)$	100	Daily water injection (m^3/d)	200

TABLE 9.4 Basic Parameters for Inverted Seven-Spot Pattern

Reservoir parameters	Parameter value	Reservoir parameters	Parameter value
Length* width* thickness of reservoir (m)	610*610*15	Volume factor (m^3/m^3)	1.18
Porosity	0.2	Total compressibility (MPa^{-1})	0.002
Oil viscosity (mpa•s)	2	Daily oil production (m^3/d)	60
Horizontal permeability $(10^{-3}\,\mu m^2)$	20	Radius of horizontal well (m)	0.08
Vertical permeability $(10^{-3}\,\mu m^2)$	20	Daily water injection (m^3/d)	300

9.4. CHAPTER SUMMARY

On the basis of the streamline numerical well testing interpretation model of a conventional vertical well, the streamline numerical well testing interpretation model of a horizontal well is studied in this chapter considering the horizontal well filtration characteristics. The main contents include:

1. The horizontal well streamline numerical interpretation model includes the filtration mathematical model of the production period and the streamline

TABLE 9.5 Basic Parameters for Inverted Nine-Spot Pattern

Reservoir parameters	Parameter value	Reservoir parameters	Parameter value
Length* width* thickness of reservoir (m)	810*810*30	Volume factor (m^3/m^3)	1.25
Porosity	0.2	Total compressibility (MPa^{-1})	0.003
Oil viscosity (mpa•s)	9	Daily oil production (m^3/d)	30
Horizontal permeability $(10^{-3}\mu m^2)$	50	Radius of horizontal well (m)	0.08
Vertical permeability $(10^{-3}\mu m^2)$	50	Daily water injection (m^3/d)	240

mathematical model of the testing period. The two mathematical models can consider not only the impact of reservoir heterogeneity, production history, oil/water phase flow, complex well pattern and complex boundary, etc., but also different horizontal well flow patterns. The filtration model of the production period is a simplified black oil model, which can simulate the production history and provide the distribution of pressure, saturation and streamline at the testing time. This model provides the basis of the establishment of the streamline numerical model in the testing period and the initial conditions of the resolution. The mathematical model of the

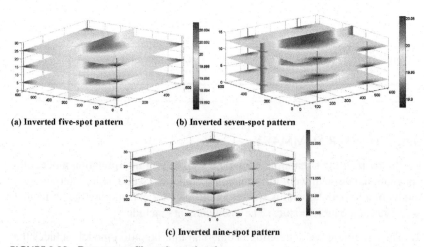

(a) Inverted five-spot pattern   (b) Inverted seven-spot pattern

(c) Inverted nine-spot pattern

FIGURE 9.22   Pressure profile at the testing time.

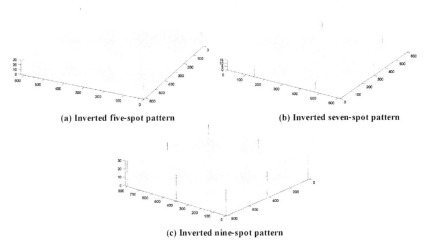

(a) Inverted five-spot pattern   (b) Inverted seven-spot pattern

(c) Inverted nine-spot pattern

FIGURE 9.23   Streamline distribution at the testing time.

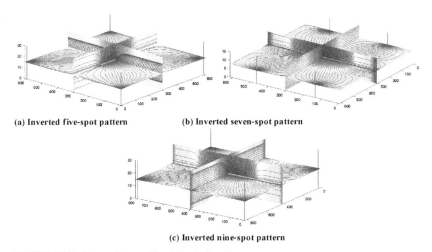

(a) Inverted five-spot pattern   (b) Inverted seven-spot pattern

(c) Inverted nine-spot pattern

FIGURE 9.24   Streamline profile at the testing time.

testing period is a streamline model, which is used to solve the pressure response of the test well. The streamline model is made up of mathematical equations established along each streamline from the test well.

2. The streamline method is used to solve the filtration mathematical model in the production period. First of all, the pressure distribution of the study area is solved by the IMPES method. Then, on the basis of the pressure distribution, the streamline tracing is taken and the saturation equation is established along each streamline. Finally, the saturation distribution at the testing time is obtained by solving the saturation equations along

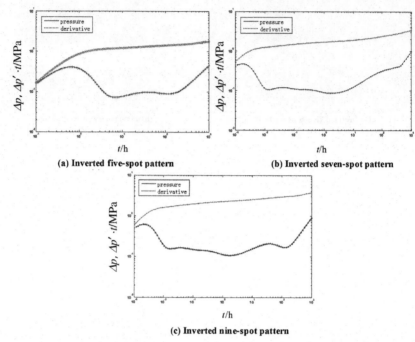

(a) Inverted five-spot pattern   (b) Inverted seven-spot pattern

(c) Inverted nine-spot pattern

FIGURE 9.25   **Pressure response curves.**

streamlines. The resolution of the streamline mathematical model in the testing period requires gathering all the streamlines from all layers which are relevant to the testing well and then solving them simultaneously.

3. By solving the typical model of the horizontal well and combining with its streamline distribution analysis, the influencing characteristics of skin factor, well bore storage coefficient, the relative thickness, the location of horizontal well, and oil/water viscosity ratio on pressure response of the streamline numerical well testing interpretation model of the horizontal well are obtained. Furthermore, the characteristics of the streamline numerical well testing interpretation model of a horizontal well in different well patterns are studied, which provides a valuable theoretical foundation for the correct interpretation of horizontal well test data.

Chapter 10

Multi-parameter Streamline Numerical Well Testing Interpretation Method

10.1. NUMERICAL WELL TESTING AUTOMATIC MATCH INTERPRETATION THEORY AND METHOD

For well testing interpretation under complex conditions, we can not combine typical curves, so only the automatic matching method can be used. The so-called automatic matching method is that using mathematical methods to make the best fit between the pressure calculated by the well testing theoretical model and the mine-measured curves. At this time, the parameters in the theoretical model are the formation parameters required. The non-linear regression analysis method is used in this paper. Compared with conventional well testing analysis methods and typical chart matching methods, this method has the following advantages:

1. it can solve the complex well testing problems with multi-parameter and variable flow rate;
2. it can evaluate the parameters correlation and parameters confidence interval, evaluating the interpretation result quantitatively;
3. it uses the system analysis method, which can avoid classifying flow pattern and make analysis results more reliable.

10.1.1. Theory of Automatic Match

Assuming $(t_i, p'_i)\,(i = 1, 2, 3, \cdots, N)$ is a group of observed pressure data. Given a group of formation parameters $\vec{x}$ (column vector), we can get a group of theoretical pressure data $(t_i, p_i)(i = 1, 2, 3, \cdots, N)$ through the well testing interpretation model. The task of well testing interpretation is to minimize the following object function by choosing a group of appropriate parameters $\vec{x}$:

$$F\left(\vec{x}\right) = \sum_{i=1}^{N} \left[p'_i - p_i\right]^2 = \sum_{i=1}^{N} f_i^2\left(\vec{x}\right)$$

That is, to get the best fit between observed pressure curve and theoretical type curve.

Streamline Numerical Well Test Interpretation. DOI: 10.1016/B978-0-12-386027-9.00010-1
Copyright © 2011 by Elsevier Ltd

10.1.2. Modified Least Square Method
10.1.2.1. Calculation Procedures

The problem was solved by using a non-linear regression analysis method to realize the goal. Take four parameters, for example, the detailed calculation procedure is as follows:

(1) Given the initial value $\vec{x}^0$, β_0, γ, ϵ, $\beta = \beta_0$, iterative index $k = 0$, $\vec{x}$;

(2) $\beta = \beta/\gamma$;

(3) Calculate $f(\vec{x}^k) = [f_1(\vec{x}), f_2(\vec{x}), \cdots, f_N(\vec{x})]^T$, $F(\vec{x}) = \sum_{i=1}^{N} f_i^2(\vec{x})$

$$f'(\vec{x}^k) = \begin{bmatrix} \dfrac{\partial f_1}{\partial x_1} & \dfrac{\partial f_1}{\partial x_2} & \dfrac{\partial f_1}{\partial x_3} & \dfrac{\partial f_1}{\partial x_4} \\ \dfrac{\partial f_2}{\partial x_1} & \dfrac{\partial f_2}{\partial x_2} & \dfrac{\partial f_2}{\partial x_3} & \dfrac{\partial f_2}{\partial x_4} \\ \dfrac{\partial f_N}{\partial x_1} & \dfrac{\partial f_N}{\partial x_2} & \dfrac{\partial f_N}{\partial x_3} & \dfrac{\partial f_N}{\partial x_4} \end{bmatrix}$$

(4) $\vec{p}^k = \left[\beta I + f'(\vec{x}^k)^T f'(\vec{x}^k)\right]^{-1} \left[f'(\vec{x}^k)^T f'(\vec{x}^k)\right]$

$$\vec{x}^{k+1} = (\vec{x}^k) + (\vec{p}^k)$$

$$F(\vec{x}^{k+1}) = \sum_{i=1}^{N} f_i^{k+1}(\vec{x}^{k+1})$$

(5) If $F(\vec{x}^{k+1}) < F(\vec{x}^k)$, then go to (7), else set $\beta = \beta\gamma$, go to (6);

(6) If $\left\| f'(\vec{x}^k)^T f'(\vec{x}^k) \right\| \leq \epsilon$, then get the optimal solution $\vec{x}^* = \vec{x}^k$;

(7) If $\left\| f'(\vec{x}^k)^T f(\vec{x}^k) \right\| \leq \epsilon$, then get the optimal solution $\vec{x}^* = \vec{x}^{k+1}$, else $k = k + 1$, go to (2).

10.1.2.2. Constraint Condition and Processing Method
10.1.2.2.1. Transformation of Constraints

Because the solved parameters have specific physical significance and explicit value ranges, the above-mentioned problem has some constraints. The interior point method is applied to get the unconstrained object function in the actual calculation, which involves constrained conditions so that the problem can be transformed into an unconstrained problem, and then the optimal solution can be calculated. The optimal solution obtained by this method is inclined to the optimal solution in the original problem and satisfies the constraints in the original problem at the same time. After transformation to an unconstrained problem, the algorithm above can be easily used to solve the problem.

The interior point method is one of the effective methods to solve the inequality constraint problem in constructing penalty function. The method's characteristic is that it can define the new constructed unconstrained object function (penalty function) in the feasible region, which can get the penalty function's minimum point.

The constraint condition in this research is:

$$x_i^{min} \leq x_i \leq x_i^{max} \quad (i = 1, 2, 3, 4)$$

The constraint condition above can be written as:

$$\begin{aligned} x_i - x_i^{min} \geq 0 \\ x_i^{max} - x_i \geq 0 \end{aligned} \quad (i = 1, 2, 3, 4)$$

The new object function constructed by the interior point method is:

$$\varphi\left(\vec{x}, \gamma^k\right) = F\left(\vec{x}\right) + \gamma^k \sum_{i=1}^{m} \left(\frac{1}{x_i^{max} - x_i} - \frac{1}{x_i - x_i^{min}} \right)$$

Where, m is the number of inequality constraints; γ' is penalty factor, a decreasing positive sequence:

$$\gamma^0 > \gamma^1 > \cdots > \gamma^k > \gamma^{k+1} > \cdots > 0$$
$$\lim_{r \to \infty} \gamma^k = 0$$

If $\vec{x}$ is in the feasible region, the value of penalty function item $\gamma^k \sum_{i=1}^{m} \left(\frac{1}{x_i^{max} - x_i} - \frac{1}{x_i - x_i^{min}} \right)$ must be positive. When $\vec{x}$ changes from constraints to the boundary $\left(x_i - x_i^{min} = 0 \text{ or } x_i^{max} - x_i = 0\right)$ in the feasible region, the value of the penalty item will sharply increase and tend towards infinity. Then the object function will also increase to infinity. Here, variable $\vec{x}$ has the penalty effect which can not go to the boundary in the iterative process, so that the constraint boundary can be regarded as a barrier for the search point not to jump out of the feasible region by the penalty item.

10.1.2.2.2. Iterative Initial Value and the Choice of Penalty Factor

As the penalty function above is defined in the parameter value range, $\vec{x}^{(0)}$ is required to satisfy all of the constraints strictly, and to avoid the boundary. Generally, the initial value should be obtained from formation parameters referencing to other methods, such as well logging or core analysis. The good initial value can speed up the iterative convergence.

The choice of initial penalty parameter (factor) $r^{(0)}$ has great influence on the calculated efficiency of the penalty function method above. If the value of $r^{(0)}$ is too small, the effect of the penalty item in the new object function (penalty function) $\varphi(\vec{x}, \gamma^k)$ will be little. Then the calculation of the unconstrained extreme in $\varphi(\vec{x}, \gamma^k)$ will be just like the calculation of unconstrained extreme in original object function $f(\vec{x})$. However, the extreme can not be close to the constrained extreme in $f(\vec{x})$ and has the probability of running out of the value range. In contrast, if the value of $r^{(0)}$ is too large, the unconstrained extreme point in the penalty function $\varphi(\vec{x}, \gamma^k)$ constructed the first few times will be far away from the constrained boundary, which can cause a decrease in the calculation efficiency.

In general, $r^{(0)}$ ranges from 1 to 10. In the calculation process, the following method is adopted: if the initial point $\vec{x}^{(0)}$ is a strict interior point, the effect of penalty item

$$\gamma^0 \sum_{i=1}^{m} \left(\frac{1}{x_i^{\max} - x_i} - \frac{1}{x_i - x_i^{\min}} \right)$$

in new object function $\varphi(\vec{x}, \gamma^k)$ should be the same as the effect of original object function $f(\vec{x}^{(0)})$. So we can derive that:

$$r^{(0)} = \left| \frac{f(\vec{x}^{(0)})}{\sum\limits_{i=1}^{mm} \left(\frac{1}{x_i^{\max} - x_i} - \frac{1}{x_i - x_i^{\min}} \right)} \right|$$

10.2. INTERPRETATION PRINCIPLE OF DOUBLE-POPULATION GENETIC ALGORITHM

10.2.1. Overall Program Design of Double-population Genetic Algorithm

The multi-parameter auto-matching process in the study of the numerical well testing interpretation method is to find a group of characteristic parameters which can describe true formation and test wells, such as permeability distribution in the control distribution of test wells, well bore storage coefficient, skin factor, etc. A genetic algorithm, which is a good random search algorithm to solve the optimization problem, is applied. Because conventional genetic algorithms are very difficult to guarantee the search speed and accuracy of optimal solution when the solution space is too big, the double-population genetic algorithm is proposed. Double-population genetic algorithm is inspired from the duty of human social division and collaboration. It meets the search speed and the accuracy of optimal solutions with the division and collaboration of the two populations at the same time: one is the overall population whose main task is to search the region where the optimal solution possibly exists (coarse search); the other is the local population whose main task is to search the region, which is defined by the overall population, to find the optimal solution with the fine search. As the algorithm uses joint search by two populations with different search strategies, it can not only ensure the accuracy of the optimal solution, but also improves the search speed. When this algorithm is applied to explain the automatic matching interpretation of numerical test, it can save computing time.

The structure of the double-population genetic algorithm is shown in Fig. 10.1.

10.2.2. Genetic Manipulation Design of Overall Population

The main task of the overall population is to search the area where the optimal solution exists. It requires large amounts of processing information, high searching speed, and it can not be limited to local minimum points, demanding low accuracy.

The overall population applies the micro-population algorithm, which has the advantage of easy calculation and high speed. It can randomly generate

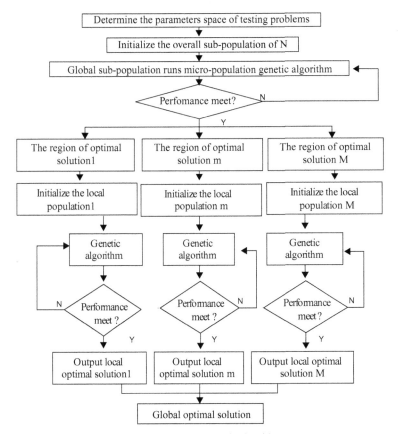

FIGURE 10.1   Structure of double-population genetic algorithm.

small populations. After some generations by genetic manipulation, the best individual can spread to the next generation, resulting in new populations. Then the genetic algorithm is repeatedly applied. The procedure is repeated until completion of the overall convergence. During the evolutionary process, new populations with constant numbers should be introduced to find the better individuals, which can ensure faster convergence speed and avoid premature convergence.

- Coding: Real number code is used.
- Generation of initial population: the solution vector is generated by the stochastic method: $\vec{X}^{(1)} = [K^{(1)}\ C^{(1)}\ S^{(1)}]^T$. Superscript 1 indicates the first population, which is the overall population.

$$K^{(1)} = K_{\min} + random\,(0, 1) \cdot (K_{\max} - K_{\min}) \qquad (10.2.1)$$

$$C^{(1)} = C_{\min} + random\,(0, 1) \cdot (C_{\max} - C_{\min}) \qquad (10.2.2)$$

$$S^{(1)} = S_{\min} + random\,(0, 1) \cdot (S_{\max} - S_{\min}) \qquad (10.2.3)$$

where, $K_{\min}, K_{\max}, C_{\min}, C_{\max}, S_{\min}, S_{\max}$ are the upper and lower limits of K, C, and S, respectively; $random\,(0,1)$ is a random number between 0 and 1.

- Assessment testing of adaptive value: the adaptive value function shows the superior and inferior of the individual or solution. In general, the adaptive value function is transformed from the objective function. If the pressure and pressure derivative data are matched at the same time, the objective function is:

$$E = a \sum_{1}^{n} [p_i - p(t_i)]^2 + b \sum_{1}^{n} \left[p_i' - p'(t_i)\right]^2 \qquad (10.2.4)$$

where, E is the objective function, $(10^{-1}\text{MPa})^2$; a is the pressure weight coefficient in the objective function; n is the ordinal number of the manometric data points; p_i is the measured pressure, 10^{-1}MPa; $p(t_i)$ is the calculated pressure at time t_i, 10^{-1}MPa; b is the pressure derivative weight coefficient in the objective function; p_i' is the measured pressure derivative, 10^{-1}MPa; $p'(t_i)$ is the calculated pressure derivative at time t_i, 10^{-1}MPa.

Some mapping transformation of the objective function range is named the scale transform of fitness. Scale transform methods commonly have: linear transformation, power function transformation and exponent transformation. Here linear transformation is used, definition of fitness function:

$$f = \begin{cases} C_m - E & E < C_m \\ 0 & E \geq C_m \end{cases} \qquad (10.2.5)$$

where, f is the fitness function, $(10^{-1}\text{MPa})^2$; C_m is constant, and its value is the maximum value during the evolution process $(10^{-1}\text{MPa})^2$.

- Selection: in accordance with the principle of survival of the fittest, sweep out of some individuals with poor fitness, while a number of new individuals are generated. Maintain the same total number of individuals. The generation method of new individuals is the same as Step 2.
- Exchange: the exchange operation of genetic algorithm is the most important genetic operation. The individuals of a new generation can be generated through the exchange operation. The new individuals combine the characteristics of their parents. A new individual is generated by selecting individuals with larger fitness value to match randomly and calculating the exchange probability.

$$\vec{X}_i^{(1)} = \alpha_1 \vec{X}_i^{(1)} + (1 - \alpha_1) \vec{X}_j^{(1)}, \quad \vec{X}_j^{(1)} = \alpha_1 \vec{X}_j^{(1)} + (1 - \alpha_1) \vec{X}_i^{(1)}$$

$$(10.2.6)$$

Where, $\vec{X}_i^{(1)}, \vec{X}_j^{(1)}$ are the individuals for exchange pairs in populations 1, $\vec{X}_i^{(1)} = [K_i^{(1)} \ C_i^{(1)} \ S_i^{(1)}]^T$, $\vec{X}_j^{(1)} = [K_j^{(1)} \ C_j^{(1)} \ S_j^{(1)}]^T$; α_1 is the exchange probability in populations 1.

- Mutation: in order to generate new individuals, individuals are randomly selected in groups $\vec{X}_m^{(1)} = [K_m^{(1)} \ C_m^{(1)} \ S_m^{(1)}]^T$, and the mutation probability is calculated. The selected individuals carry out mutation by a certain mutation probability.

$$K_m^{(1)} = (1 - \beta_K)K_m^{(1)} + \beta_K \cdot random(0, 1) \cdot (K_{max} - K_{min}) \qquad (10.2.7)$$

$$C_m^{(1)} = (1 - \beta_C)C_m^{(1)} + \beta_C \cdot random(0, 1) \cdot (C_{max} - C_{min}) \qquad (10.2.8)$$

$$S_m^{(1)} = (1 - \beta_S)S_m^{(1)} + \beta_S \cdot random(0, 1) \cdot (S_{max} - S_{min}) \qquad (10.2.9)$$

Where, $\vec{X}_m^{(1)}$ is individuals selected for mutation. β_K, β_C, and β_S are the mutation probabilities of K, C, and S, respectively.

- Optimal strategy of preservation: introducing the generation gap technology, the best individuals can directly go to the next generation without genetic manipulation.
- Output of the region of optimal solution: after certain genetic generations, the region of optimal solution is output by genetic operation.

10.2.3. Genetic Manipulation Design of Local Population

The main task of the local population is to search optimal solution in the region which has local optimal solution. The range of search is small, but it requires higher accuracy. So a different search area and encoding length from the overall population should be used.

- Coding: real number code is used.
- Generation of initial population: Solution vector is generated by stochastic method: $\vec{X}^{(2)} = [K^{(2)}\ C^{(2)}\ S^{(2)}]^T$. Superscript 2 indicates the second population, which is the local population.

$$K^{(2)} = K_i^{(1)} - R_K + random\ (0,1) \cdot (2R_K) \tag{10.2.10}$$

$$C^{(2)} = C_i^{(1)} - R_C + random\ (0,1) \cdot (2R_C) \tag{10.2.11}$$

$$S^{(2)} = S_i^{(1)} - R_S + random\ (0,1) \cdot (2R_S) \tag{10.2.12}$$

where, R_K, R_C, R_S are the search radius of the component $K^{(2)}$, $C^{(2)}$, $S^{(2)}$ respectively, which are the center component of the optimal solution $\vec{X}_i^{(1)} = [K_i^{(1)}\ C_i^{(1)}\ S_i^{(1)}]^T$ in the overall population.

- Assessment testing of adaptive value: the objective function and fitting function can use the same ones as the overall population, which is shown, respectively, in Equations 10.2.4 and 10.2.5.
- Selection: according to the principle of survival of the fittest, individuals with good adaptability have larger probability to enter the next generation. Using the roulette wheel selection method to select the superior individuals from the current individuals and sweep out of poor individuals. A number of new individuals are generated at the same time and the same total number of individuals should be maintained.
- Exchange: then select individuals with larger fitness value to match randomly and calculate the exchange probability.

$$\vec{X}_i^{(2)} = \alpha_2 \vec{X}_i^{(2)} + (1 - \alpha_2)\vec{X}_j^{(2)}, \ \vec{X}_j^{(2)} = \alpha_2 \vec{X}_j^{(2)} + (1 - \alpha_2)\vec{X}_i^{(2)} \tag{10.2.13}$$

where, $\vec{X}_i^{(2)}$, $\vec{X}_j^{(2)}$ are the individuals for exchange in populations 2, $\vec{X}_i^{(2)} = [K_i^{(2)}\ C_i^{(2)}\ S_i^{(2)}]^T$, $\vec{X}_j^{(2)} = [K_j^{(2)}\ C_j^{(2)}\ S_j^{(2)}]^T$; α_2 is the exchange probability in populations 2.

- Mutation: in order to generate new individuals, individual $\vec{X}_m^{(2)} = [K_m^{(2)}\ C_m^{(2)}\ S_m^{(2)}]^T$ is randomly selected from the population, and some string values in its string structure are randomly changed with a certain probability.

$$K_m^{(2)} = (1 - \beta_K)K_m^{(2)} + \beta_K \cdot random\ (0,1) \cdot (2R_K) \tag{10.2.14}$$

$$C_m^{(2)} = (1 - \beta_C)C_m^{(2)} + \beta_C \cdot random\ (0,1) \cdot (2R_C) \tag{10.2.15}$$

$$S_m^{(2)} = (1 - \beta_S)S_m^{(2)} + \beta_S \cdot random\ (0,1) \cdot (2R_S) \tag{10.2.16}$$

- Optimal strategy of preservation: introducing the generation gap technology, the most fit individual of each generation can unconditionally go into the next generation without genetic manipulation.
- Terminate condition: the objective function E is examined to see whether it can meet the requirements. If it can satisfy the requirements, the genetic algorithm will terminate and output the optimal solution. Otherwise, the genetic algorithm will continue.

The double-population genetic algorithm combines global and local searching to carry out rough searching in the overall solution space and fine searching in the local solution space. It not only has a higher degree of automation than the conventional genetic algorithm, but also avoids running into the local optimal solution, and ensures the accuracy of the optimal solution. Application results show that the double-population genetic algorithm is faster than conventional genetic algorithms under the same search solution space and accuracy requirement of the optimal solution, which has higher application value in the automatic fitting interpretation of numerical well testing.

10.3. CHAPTER SUMMARY

Based on the streamline numerical well testing theory above, the multi-parameter auto-matching interpretation method of streamline numerical well testing, which uses a double-population genetic algorithm, is studied in this chapter. There are two populations used in the searching course: the global population and the local population. The velocity of solution searching is enhanced and the accuracy of solution is ensured by combining two populations.

Chapter 11

Software Programming of Streamline Numerical Well Testing Interpretation

11.1. OVERVIEW

11.1.1. Design Background

The well testing method has become an important dynamic monitoring method in the oil/gas field development. With the oil field entering the later stage of water flooding and the development of some special types of oil/gas field, to a large extent, the current well testing theory and interpretation methods can not meet the real situation of field underground filtration during medium-high water cut. The main problem is that the well testing interpretation model is too simple, and does not consider the impact of multi-phase fluid flow, multi-well interference, production history and multi-layering. In order to meet the needs of real production, the complex well testing interpretation model considering multi-geological factors and development factors was established on the basis of the basic well testing theory and then solved by the streamline method, which forms a complete numerical well testing theory and interpretation method. The streamline well testing interpretation model considering a number of factors was established first, and the numerical well testing simulator, which could consider factors such as reservoir heterogeneity, development well pattern, production history and multi-phase fluid, was formed. Practical "Streamline Numerical Well Testing Interpretation Software" was developed.

11.1.2. Main Performances

Based on the application of the latest reservoir engineering streamline methods and global optimization algorithm–genetic algorithm, "Streamline Numerical Well Testing Interpretation Software" was developed with Visual C++ 6.0 programing language and Matlab 6.5 mathematical language in the Microsoft Windows® operating system. The object-oriented programing method is achieved and the current advanced programing techniques, such as procedure plug-in link, are comprehensively used. The distinguished features are convenient, amicable, beautiful and practical, and so on. All operations of software are

Streamline Numerical Well Test Interpretation. DOI: 10.1016/B978-0-12-386027-9.00011-3
Copyright © 2011 by Elsevier Ltd

menu-based, so it is easy to operate. The software has a strong self-protection function and better misinterpretation- and error-handling abilities. It can automatically identify illegal information and give tips to users. The software development team strictly accorded with the idea of software engineering in the process of programing. The program has four main function modules (geological model module, production history calculation module, well testing interpretation module, graphic display module), and each module can complete its function independently. The data between modules are transferred by citing data files, so it is easy to modify and maintain.

11.1.3. Main Features

The software project is on the basis of the basic reservoir engineering theory, which has applied the latest streamline method and the improved genetic algorithm to numerical well testing interpretation. Through automatic matching interpretation of the pressure data from testing wells, well testing parameters solution and the remaining oil distribution, etc., can be obtained from the software. Compared with other peer software, it has the following significant features: comprehensive factors considered (production history, combined production and injection in multi-layer, multi-phase flows, multi-well interference, heterogeneity, complex outer boundary, displacement ways, etc.), high calculation speed (two-dimensional or three-dimensional problems will be transformed into one-dimensional problems along streamlines by the streamline method), rich interpretation parameters (besides the parameters of conventional well testing interpretation, remaining oil distribution, polymer concentration distribution are also included).

11.2. INTRODUCTION TO SOFTWARE FUNCTIONS

The overall design of the "Streamline Numerical Well Testing Interpretation Software" is shown in Fig. 11.1.

The "Streamline Numerical Well Testing Interpretation Software" consists of four main function modules, as listed below.

11.2.1. Geological Modeling Module

The basic functions are as follows:

1. well position is automatically divided and boundary is automatically treated;
2. the grid discrete parameters are automatically generated, including the effective thickness of reservoir and sand thickness distribution, permeability distribution and porosity distribution, etc.;
3. initial conditions files are automatically generated;
4. outer boundary conditions are automatically generated;
5. physical parameters files are automatically generated;

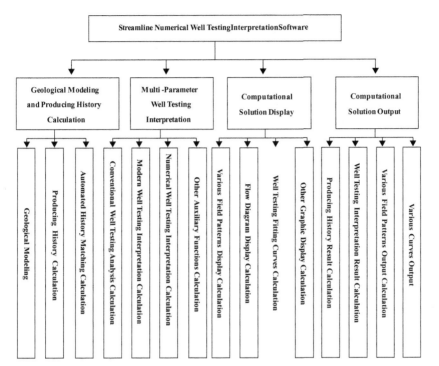

FIGURE 11.1   Overall design of streamline numerical well testing interpretation software.

6. control parameters files are automatically generated; and
7. graphical display and output, and other features.

11.2.2. Calculation Module of Production History

The primary function of this module is to solve the mathematical model of the filtering flow during production, so as to obtain the well testing parameters such as pressure and saturation distribution at the test time. The basic functions of the module are as follows:

1. production history calculation of water flooding/polymer flooding/combination flooding;
2. the view of the production indicators matching situation, including the water-cut of single-well and the blocks' comprehensive water-cut fitting curve;
3. manual accommodation of history matching parameters;
4. production dynamic contrast, which can compare production historical data of multi-wells by months;
5. automatic history matching of production history calculation; and
6. the format settings, output and other features, of above graphics.

11.2.3. Graphics Display Module

The basic functions are as follows:

1. display of porosity parameter in three modes;
2. display of sand thickness parameter in three modes;
3. display of permeability parameter in three modes;
4. display of effective thickness parameter in three modes;
5. display of formation pressure distribution at various stages in three modes;
6. display of oil saturation distribution at various stages in four modes;
7. display of geological reserves abundance distribution at various stages in two modes;
8. display of polymer concentration distribution at various stages in three modes;
9. display of alkali concentration distribution at various stages in three modes;
11. streamlined distribution display at various stages;
11. graphic display of overall development indicators in the blocks, including the number of open oil/water wells, monthly production/injection data, monthly injection–production ratio, accumulated production/injection data, cumulative injection–production ratio, composite water cut; and
12. format settings and output of above graphics and other features.

11.2.4. Well Testing Interpretation Module

The basic functions of this module are as follows:

1. pretreatment of well testing data, such as inputting the basic information of testing wells, the format setting of piezometric data, data vacuate and smoothing process, end-point processing of well testing data (extraction of the test section), calibration of the shut-in pressure and so on;
2. check the test curve in three cases, including rectangular coordinates, semi-log coordinates and log–log coordinates;
3. check the theoretical curve and the characteristics of the theoretical curve under given testing parameters of testing wells (including the complex factors, such as heterogeneity, two-phase flow, multi-layer);
4. numerical well testing analysis, including manually matching analysis and automatically matching analysis;
5. conventional (semi-logarithmic) well testing analysis, including the drawdown analysis and build-up test analysis;
6. modern well testing analysis, including manually matching analysis and automatically matching analysis;
7. reservoir delineation test; and
8. other functions.

11.3. CHAPTER SUMMARY

In the Microsoft Windows® operating system, "Streamline Numerical Well Testing Interpretation Software" was compiled with Visual C++ 6.0 programing language and Matlab 6.5 mathematical language, which is convenient, amicable, beautiful, practical and easy to operate. What is more, it is self-protecting with regard to misinterpretation and mishandling. This software is divided into four main function modules, and each module can complete its function independently. The data between modules are transferred by citing data files, so it is easy to modify and maintain.

Field Application of Streamline Numerical Well Testing Interpretation Software

The improved system of interpretation theory on streamline numerical well testing has been formed after nearly a decade of development of the streamline numerical well testing interpretation method. The well test interpretation software with independent intellectual property rights (registration number of the software copyright: 2008SR00065) has been developed. More than 20 blocks, such as Shengli Oilfield, Nanyang Oilfield, Zhongyuan Oilfield, Dagang Oilfield have been popularized and applied. Three examples will be systematically introduced in this chapter.

12.1. APPLICATION CASE ONE

12.1.1. Survey of Fault Block Geology and Development

The block layer of this study is a fan-shaped delta deposit in the middle of a fan-delta. The layer of reservoir sediment is thick and its lithology is mainly rudaceous sandstone and rough pebbly sandstone. The oil area of layer is $1.0 \, \text{km}^2$, geological reserve is $220 \times 10^4 \text{t}$, formation temperature is $70\,^{\circ}\text{C}$, and the crude oil viscosity is $10.0 \, \text{mPa} \bullet \text{s}$. There are five layers in total, the largest of which is $1.0 \, \text{km}^2$, and the smallest of which is $0.5 \, \text{km}^2$. The layers have good superimposable properties in the vertical. The average porosity is 21.3% and the average air permeability is $0.628 \, \mu\text{m}^2$. The interlayer coefficient of permeability variation is $0.35{\sim}0.6$, the intraformational coefficient of permeability variation is $0.5{\sim}0.79$, the average effective thickness is $16.2 \, \text{m}$, the initial pressure is $14.5 \, \text{MPa}$, the initial water saturation is 0.3.

The block has been producing for nearly 30 years, and has experienced six stages of development: commingled water injection and production, subdivision adjustment, first and second infill adjustment of well pattern, local completeness adjustment of well pattern and polymer flooding, etc.

In order to study the residual oil distribution after polymer flooding, six wells are selected to build-up testing and drawdown testing in one month, which aims

Streamline Numerical Well Test Interpretation. DOI: 10.1016/B978-0-12-386027-9.00012-5
Copyright © 2011 by Elsevier Ltd

to quantitatively describe the remaining oil distribution in the block using these well testing data.

12.1.2. Streamline Numerical Well Testing Interpretation

The model has five layers and adopts a uniform grid system in the plane. There are 36 grids in the X direction and its grid step is 60 m, 16 grids in the Y direction and its grid step is 60 m, and 5 grids in the Z direction which are non-uniform grids, the grid step is the effective thickness. The total number of grid nodes is $36 \times 16 \times 5 = 2880$. Kriging interpolation has been used to obtain the effective thickness distribution and permeability distribution of each layer. There are 17 production wells and 13 injection wells in the block historically.

Based on this reservoir model, the water-cut curve of block and single well (Figs 12.1 and 12.2), the streamline curve of the block at shut-in time (Figs 12.3 to 12.7), the pressure matching curve of six testing wells (Figs 12.8 to 12.13; the interpretation results are shown in Table 12.1), permeability distribution (Figs 12.14 to 12.18), and the remaining oil saturation distribution (Figs 12.19 to 12.23) are ultimately obtained after matching water cut of the block during production phase and matching data from six test wells during the testing phase.

12.2. APPLICATION CASE TWO

12.2.1. Survey of Block Geology and Development

The object studied in application case two is the block reservoir of fine-middle coarse lithic arkose in neogene guantao formation. Guantao formation reservoir is divided into four oil groups according to the differences in sedimentary

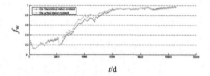

FIGURE 12.1   Composite water-cut matching curve of the block.

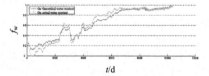

FIGURE 12.2   Water-cut matching curve of P1.

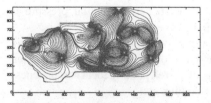

FIGURE 12.3   Streamline distribution of L1 at shut-in test time.

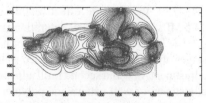

FIGURE 12.4   Streamline distribution of L2 at shut-in test time.

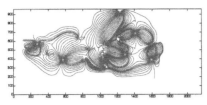

FIGURE 12.5   Streamline distribution of L3 at shut-in test time.

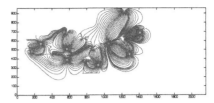

FIGURE 12.6   Streamline distribution of L4 at shut-in test time.

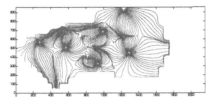

FIGURE 12.7   Streamline distribution of L5 layer in the block at shut-in test time.

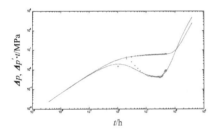

FIGURE 12.8   Log-log graph on well testing pressure matching of P6.

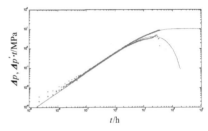

FIGURE 12.9   Log-log graph on well testing pressure matching of P7.

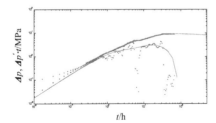

FIGURE 12.10   Log-log graph on well testing pressure matching of I4.

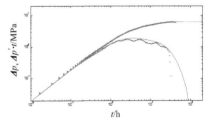

FIGURE 12.11   Log-log graph on well testing pressure matching of I7.

cycle characteristics, oil layer development characteristics, claystone interlayer development degree and crude oil properties, which are AI, upper AII, lower AII and AIII oil group, respectively, from top to bottom. Oil area of AII up group is $4.5\,\mathrm{km}^2$, mean effective thickness of the oil layer is $12.3\,\mathrm{m}$, geological reserve is $938 \times 10^4\,\mathrm{t}$, reservoir temperature is $63\,^\circ\mathrm{C}$, initial water

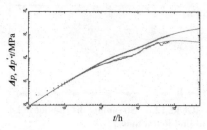

FIGURE 12.12 Log-log graph on well testing pressure matching of I8 well.

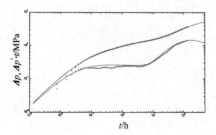

FIGURE 12.13 Log-log graph on well testing pressure matching of I9 well.

TABLE 12.1 Data Sheet of Test Well Interpretation for Application Case One

Well	Parameter										
	Permeability of well point (μm^2)					Skin factor (dimensionless)					Well bore storage factor (m^3 / MPa)
	L1	L2	L3	L4	L5	L1	L2	L3	L4	L5	
P6	0.51	0.37	/	0.44	1.12	−0.4	3.1	/	−0.3	0.1	0.228
P7	0.42	0.43	0.15	0.83	0.78	−0.3	−2.1	−0.1	−0.3	−0.2	1.120
I4	/	/	/	0.86	0.82	/	/	/	−1.2	0.1	0.125
I7	0.26	0.28	0.66	/	/	−0.5	−0.2	−0.3	/	/	0.231
I8	0.19	0.78	/	/	0.82	0.2	0.3	/	/	0.1	0.163
I9	0.58	0.26	0.08	/	/	−0.2	−0.2	−1.3	/	/	0.023

saturation is 0.35, initial pressure is 13.32 MPa, crude oil density in the stock tank is 0.96g/ cm³, underground crude oil viscosity is 100 mPa•s, crude oil volume factor is 1.1, crude oil compressibility factor is 7.2×10^{-4}MPa^{-1}, formation water density is 1.0 g/cm³, formation water compressibility coefficient is 6.75×10^{-4}MPa^{-1}, formation water viscosity is 0.45 mPa•s, reservoir buried depth is 1316.8–1396.0m, and four sublayers are divided from top to bottom(1, 2, 3 and 4). Sublayers 3 and 4 have the characteristics of large

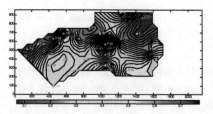

FIGURE 12.14 Permeability distribution of L1 (μm^2).

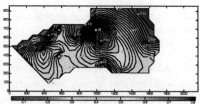

FIGURE 12.15 Permeability distribution of L2 (μm^2).

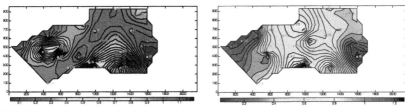

FIGURE 12.16 Permeability distribution of L3 (μm^2).

FIGURE 12.17 Permeability distribution of L4 (μm^2).

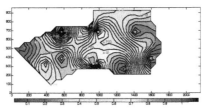

FIGURE 12.18 Permeability distribution of L5 (μm^2).

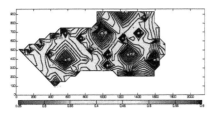

FIGURE 12.19 Oil saturation distribution of the L1 during the testing phase.

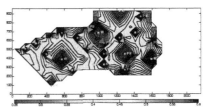

FIGURE 12.20 Oil saturation distribution of the L2 during the testing phase.

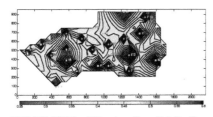

FIGURE 12.21 Oil saturation distribution of the L3 during the testing phase.

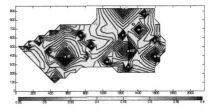

FIGURE 12.22 Oil saturation distribution of the L4 during the testing phase.

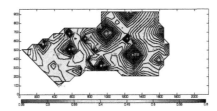

FIGURE 12.23 Oil saturation distribution of the L5 during the testing phase.

thickness and wide distribution, and they are the major layers of AII upper oil group which is the research target here.

The block has been producing for nearly 30 years, and has experienced six stages of development: drilling development well pattern, water flooding, the first adjustment, the second adjustment, the third adjustment, and alkaline–polymer combination flooding. There are historically 121 production wells and 56 injection wells.

This block enters the later development period with super-high water cut, and reservoir physical properties and fluid distribution are greatly influenced by long-term water injection. In the dominant flowing path of the high permeability zone, effective water injection weakens. In order to study the fluid pattern modification and profile control, distribution zone and relevant parameters of dominant flowing path are quantitatively analyzed. In another way, well testing data are used to describe residual oil variation rule between the injection and the production well, alkaline/polymer front position and mobile oil saturation variation rule quantitatively. Also, underground residual oil distribution rule and development characters are totally recognized after polymer flooding, and this can provide the basis for improving development effect after polymer flooding and further developing underground residual resources. For one month, 24 wells were chosen for pressure build-up and drawdown test, including 13 injection wells, three polymer injection wells, two wells with polymer injection history, and six oil wells.

Block structure well location of case two is shown in Fig. 12.24.

12.2.2. Streamline Numerical Well Testing Interpretation

The model is a two-layer reservoir. A uniform grid system is used in the plane. There are 51 grids in direction X and its grid step is 50 m; there are 71 grids in direction Y and its grid step is 50 m, and there are two grids in direction Z, which are non-uniform grids and grid step is sand thickness. The total number of grid nodes is $51 \times 71 \times 2 = 7242$. There is edge water in the northeast and southeast of the target reservoir, and the other directions are closed fault; grid division and boundary treatment are shown in Fig. 12.25.

Kriging interpolation is used to obtain sand thickness distribution (Figs 12.26 to 12.27), effective thickness distribution (Figs 12.28 and 12.29), permeability distribution (Figs 12.30 and 12.31) and porosity distribution (Figs 12.32 and 12.33) of each layer.

In order to achieve agreement between the built geological model and the real reservoir, effective thickness and porosity distribution are slightly adjusted to fit geological reserve of upper AII layer 3 and 4. Geological reserve of upper AII layer 3 is 707.61×10^4 t, fitting geological reserve is 726.54×10^4 t, fitting relative error is 2.67%; geological reserve of upper AII layer 4 is 193.69×10^4 t, fitting geological reserve is 205.01×10^4 t, fitting relative error is 5.84%; total geological reserve of upper AII layer 3 and 4 is 901.3×10^4 t, fitting geological

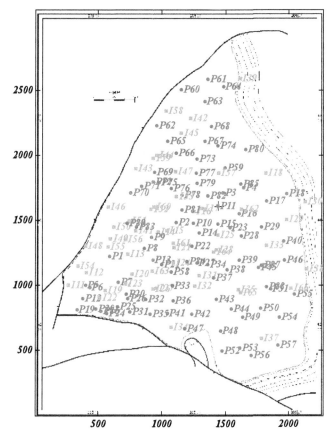

FIGURE 12.24   Block structure well location of application case two.

reserve is 931.6×10^4 t, fitting relative error is 3.36%. Based on this reservoir model, water cut of block and most production wells are fitted in the production period, and 24 testing wells data are fitted in the testing period. The following curves at shut-in time are obtained: block and single-well water fitting curve in the testing period (Figs 12.34 to 12.41), oil saturation distribution (Figs 12.42 and 12.43), pressure distribution (Figs 12.44 and 12.45), streamline distribution (Figs 12.46 and 12.47), polymer concentration distribution (Figs 12.48 and 12.49) and alkaline distribution (Figs 12.50 and 12.51).

12.2.2.1. Well Testing Interpretation of the Production Well
12.2.2.1.1. Well Testing Interpretation of Well P35

- The testing period of well P35 is from 2006.7.7 to 2006.7.10, and the testing interval is 1351.6–1390 m; testing formations are upper AII layer 3 and upper

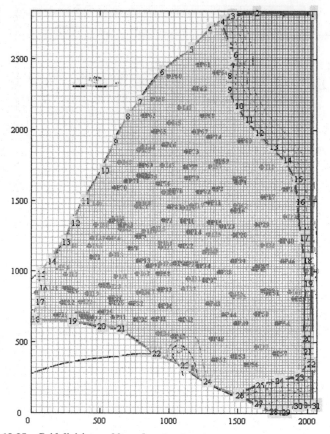

FIGURE 12.25   Grid division and boundary treatment.

AII layer 4. The production dynamic curve is shown in Fig. 12.52, and well location is shown in Fig. 12.53.

- The testing curve conformation analysis of well P35. The testing curve of this well is soft and can be divided into early, middle and late periods. The early period is influenced by well bore storage effect; pressure and its derivation curve moves upwards with unit slope. The middle period is a radial flow straight-line period; the pressure derivation curve shows a horizontal straight segment; after a short radial flow period, fluid flows experiences late period, the pressure response curve of the late period is influenced by southern block, and the pressure derivation curve shows an upwarp. The pressure response fitting curve of this well is shown in the Fig. 12.54. It can be seen that the fitting result is very good and the interpretation results are shown in Table 12.2.
- Summary of the interpretation results. The detected fault range of this well basically agrees with the geological structure; the pressure derivation response curve of the testing well begins upwarp at 14.6 h; and the

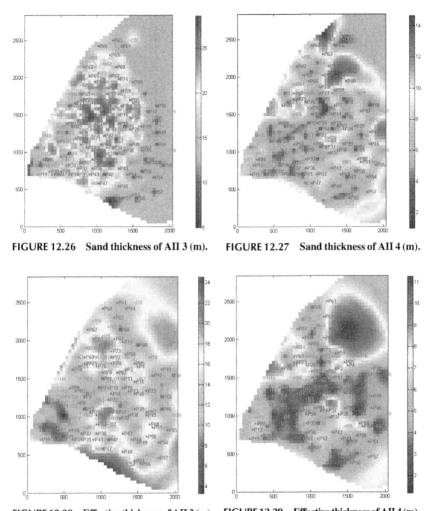

FIGURE 12.26   Sand thickness of AII 3 (m).   FIGURE 12.27   Sand thickness of AII 4 (m).

FIGURE 12.28   Effective thickness of AII 3 (m).   FIGURE 12.29   Effective thickness of AII 4 (m).

calculated fault range is 147 m. It can be seen from seismic data that a fault exists 150 m south of the testing well, and the pressure derivation curve shows upwarp in the late period due to the influence of the southern fault.

12.2.2.1.2. Well Testing Interpretation of Well P85

- Testing survey of well P85. The testing period was from 2006.07.28 to 2006.07.31; the testing well segment was 1360.0–1394.5 m; the testing layer was upper AII 3; the production performance curve is shown in Fig. 12.55, and well location is shown in Fig. 12.56.
- Testing curve conformation analysis of well P85. The pressure and its derivation curve in this well show a rising tendency in the early period; the

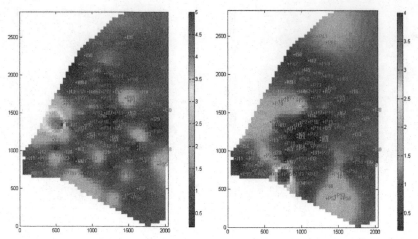

FIGURE 12.30   Permeability distribution of AII 3 (μm^2).

FIGURE 12.31   Permeability distribution of AII 4 (μm^2).

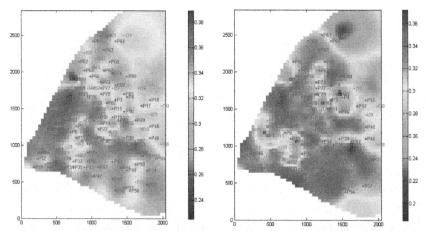

FIGURE 12.32   Porosity distribution of AII 3.

FIGURE 12.33   Porosity distribution of AII 4.

slope is small (about 1/2) and the maximum value of the pressure variation is less than 1MPa, which indicates that permeability is good around the well and the oil well decompresses quickly. In the middle period, the pressure derivation shows an upwarp; analysis shows that this is because well P85 is near polymer injection well P3; pressure response enters polymer slug region after testing well shut-in, and then fluid mobility decreases (maximum upwarp point is the polymer flooding front). In the late period, the pressure derivation curve declines continuously. The pressure response fitting curve of this well is seen in Fig. 12.57; fitting curve conformation agrees with the variation rule of the real measured curve; the interpretation result is shown in Table 12.3.

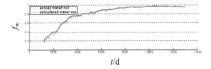

FIGURE 12.34   Fitting curve of block composite water cut.

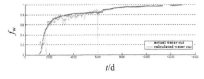

FIGURE 12.35   Fitting curve of water cut of P15.

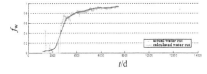

FIGURE 12.36   Fitting curve of water cut of P27.

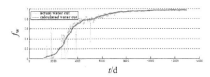

FIGURE 12.37   Fitting curve of water cut of P35.

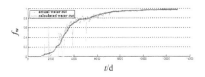

FIGURE 12.38   Fitting curve of water cut of P39.

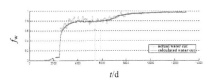

FIGURE 12.39   Fitting curve of water cut of P42.

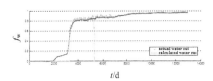

FIGURE 12.40   Fitting curve of water cut of P54.

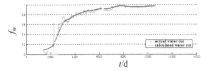

FIGURE 12.41   Fitting curve of water cut of well P72.

- Summary of interpretation results. Until the shut-in testing time of well P85, well P3 in its southwest 215 m, cumulative polymer injection is 122.014 t, alkaline injection is 331.445 t, and cumulative solution injection is $12.7 \times 10^4 \, m^3$. It is predicted from the material balance method of reservoir engineering that the displacement front of the combination flooding has moved 76 m forward and has reached 139 m away from well P85, which agrees with the investigation radius result 121 m after shut-in of 6.81 h.

12.2.2.1.3. Well Testing Interpretation of Well P81

- Testing survey of well P81. Testing period was 2006.7.24 to 2006.7.27; testing interval was 1345.3–1379 m; and testing formations were upper AII layer 3 and upper AII layer 4. The production dynamic curve is shown in Fig. 12.58, and well location is shown in Fig. 12.59.

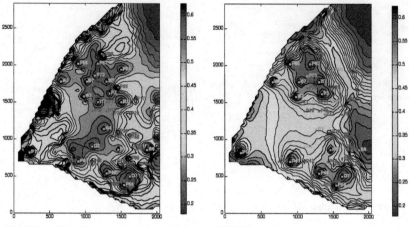

FIGURE 12.42   Oil saturation distribution of AII 3.

FIGURE 12.43   Oil saturation distribution of AII 4.

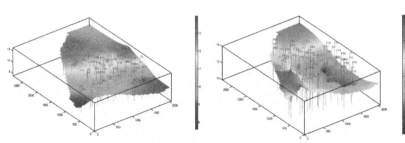

FIGURE 12.44   Pressure distribution of AII 3 (MPa).

FIGURE 12.45   Pressure distribution of AII 4 (MPa).

- Testing curve conformation analysis of well P81. The testing curve of this well is divided into early, middle and late periods. The early stage is influenced by well storage effect, which is very weak for the small well storage coefficient and middle radial flow appears quickly. In the middle radial flow period, the pressure derivation curve fluctuates slightly, indicating that formation mobility has little variation; after the radial flow period, fluid flows experiences late period, pressure derivation curve shows upwarp (analysis shows that this is because well P81 is near polymer injection well I17), pressure response enters polymer slug zone after well shut-in and the fluid mobility decreases (the peak point of upwarp is the polymer flooding front). Figure 12.60 shows the pressure response fitting curve of this well; fitting curve conformation agrees with the variation rule of the measured curve, and interpretation results are shown in Table 12.4.
- Summary of interpretation results. Until shut-in testing time of well P81, well I17 in its east 215 m, cumulative polymer injection is 121.911 t, alkaline injection is 331.591 t, and cumulative solution injection is $12.6 \times 10^4\, m^3$.

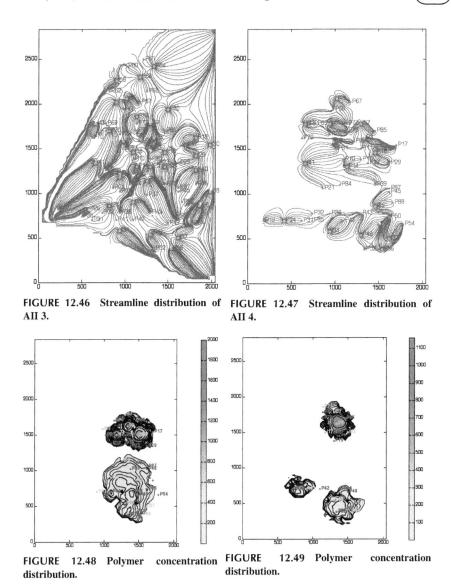

FIGURE 12.46 Streamline distribution of AII 3.

FIGURE 12.47 Streamline distribution of AII 4.

FIGURE 12.48 Polymer concentration distribution.

FIGURE 12.49 Polymer concentration distribution.

It is evaluated from the material balance method of reservoir engineering that the displacement front of the combination flooding has moved 79 m forward and reached a point 121 m away from well P81, which agrees with the polymer injection front result 132 m after shut-in of 2.45 h.

12.2.2.1.4. Well Testing Interpretation of Well P38

- Testing survey of well P38. The testing period was 2006.7.18 to 7.22; the testing interval was 1343.3–1379.8 m; and testing formations were upper AII

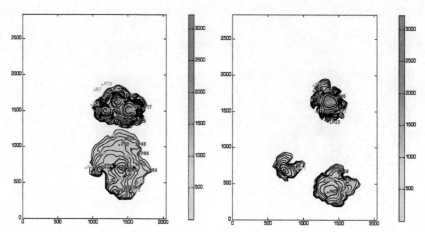

FIGURE 12.50 Alkaline concentration distribution of AII 3 (mg/l).

FIGURE 12.51 Alkaline concentration distribution of AII 4 (mg/l).

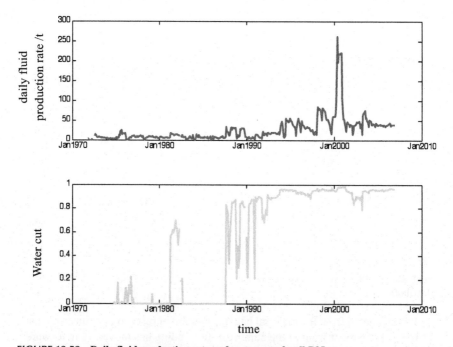

FIGURE 12.52 Daily fluid production rate and water cut of well P35.

layer 3 and upper AII layer 4. The production dynamic curve is shown in Fig. 12.61, and well location is shown in Fig. 12.62.
- Testing curve conformation analysis of well P38. The pressure response curve of this well in the early and middle periods shows a rising tendency; the slope is small (about 1 / 2) and the maximum value of pressure variation is

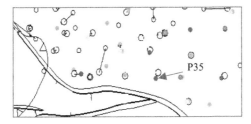

FIGURE 12.53   Well location of P35.

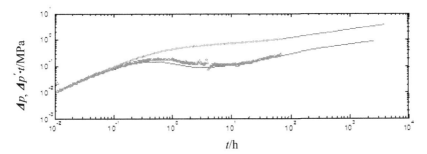

FIGURE 12.54   Pressure response fitting curve of P35.

TABLE 12.2 Well Testing Interpretation Result of Well P35

Permeability (μm^2)		Skin factor (dimensionless)		Well bore storage factor (m^3 / MPa)	Fault range (m)	
					(T = 14.16 h)	
All3	All4	All3	All4		All3	All4
0.68	1.44	−1.0	−2.0	0.4	147	171

less than 1MPa, which indicates that permeability is good around the well and the oil well decompresses quickly. In the later part of middle period, the pressure derivation curve shows slight fluctuation; considering that well I65 around this well is a former polymer injection well, this could indicate that the fluctuation is caused by mobility variation of residual polymer in the testing well control region. In the late period, the pressure derivation curve drops, and this is due to continuous production interference of surrounding production wells at well shut-in testing time. The pressure response fitting curve of this well is shown in Fig. 12.63, and the fitting curve conformation agrees with the variation rule of the measured curve. The interpretation results are shown in Table 12.5.

- Summary of interpretation results. After well P38 shut-in of 2.86 h, the pressure derivation curve shows slight fluctuation. According to the

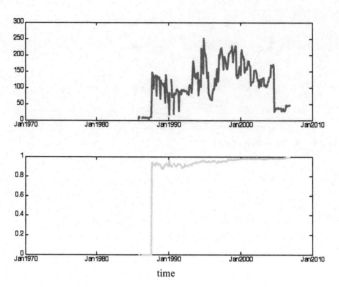

FIGURE 12.55   Daily fluid production and water cut of well P85.

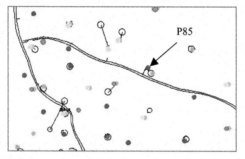

FIGURE 12.56   Well location of well P85.

specific conditions of this region, this fluctuation is caused by mobility variation of residual polymer in the testing well control region. According to investigation radius of upwarp peak point, residual polymer is located in 85 m away from the testing well.

12.2.2.2. Well Testing Interpretation of a Polymer Injection Well
12.2.2.2.1. Well Testing Interpretation of Wells I17 and I62

- Survey of wells I17 and I62. Wells I17 and I62 are polymer injection wells. Polymer and alkaline were injected into well I17 simultaneously (binary compound flooding) in May 2005; polymer was injected into well I62 in May 2005 and alkaline (binary compound flooding) was injected into well

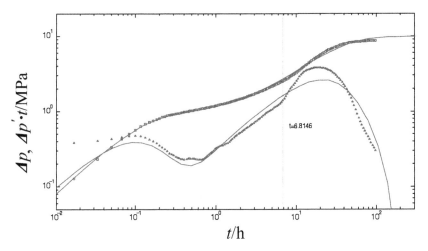

FIGURE 12.57   Pressure response fitting curve of well P85.

TABLE 12.3 Well Testing Interpretation Result of Well P85

Permeability (μm²)		Skin factor (dimensionless)		Well bore storage factor (m³ / MPa)	Polymer injection front range (m)	
					(T = 6.81 h)	
AII3	AII4	AII3	AII4		AII3	AII 4
1.43	/	−3.0	/	0.25	121	/

I62 in January 2006. The testing period of the two wells was 2006.6.22 to 2006.7.6; the testing interval was 1343.3–1379.8 m; and both the testing formations of well I17 and I62 are upper AII layer 3. The production dynamic curves of wells I17 and I62 are shown in Fig. 12.64 and Fig. 12.65, respectively, and well location is shown in Fig. 12.66.

- Testing curve conformation analysis of well I17 and I62. The pressure response curve conformation of wells I17 and I62 are similar: pressure and its derivation curves are two parallel straight lines showing slowly rising tendencies, which are similar to the characteristics of fracture formation. This is due to the existence of dominant flowing path in formation; at the same time, because of polymer injection, mobility deteriorates and the non-Newtonian effect appears; the pressure derivation curve drops in the late period, this variation is caused by that, the pressure response enters the oil zone after polymer slug zone and total mobility increases, and the lowest point is the polymer front. The pressure derivation curve shows upwarp at the end, and this is the second active interference of surrounding active well P1 in

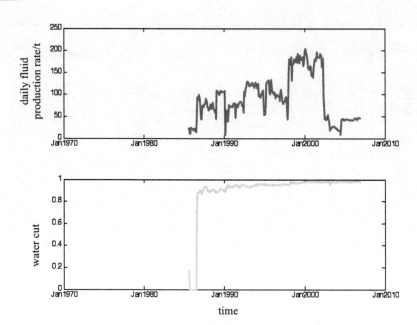

FIGURE 12.58   **Daily fluid production rate and water cut of well P81.**

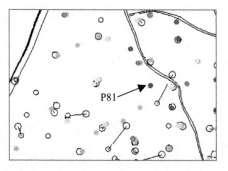

FIGURE 12.59   **Well location of well P81.**

late production. Fitting curve shape of this well (Figs 12.67 and 12.68) agrees with variation rule of the measured curve, and the interpretation result is shown in Table 12.6.

- Summary of testing results. For well I17, by the well shut-in time, cumulative polymer injection is 116.848 t, alkaline injection is 312.386 t, and cumulative solution injection is $12.2 \times 10^4 \, m^3$. It is evaluated from the material balance method of reservoir engineering that displacement front of combination flooding moves 78 m forward and agrees with calculation result 74 m of polymer injection front range after well shut-in of 179 h, which verified the analysis result of well testing. The pressure derivation curve drops in

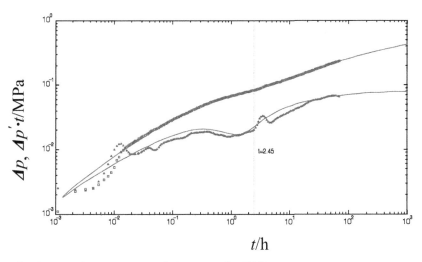

FIGURE 12.60   Pressure response fitting curve of well P81.

TABLE 12.4 Well Testing Interpretation Result of Well P81

Permeability (μm²)		Skin factor (dimensionless)		Well bore storage factor (m³ / MPa)	Polymer injection front range (m)	
					(T = 2.45 h)	
All3	All4	All3	All4		All3	All4
1.46	0.87	−3.0	−2.0	0.4	183	132

the late period, pressure response enters the oil zone after polymer slug zone and total mobility increases, which can judge polymer flooding front (the lowest point of pressure derivation).

- The situation of well I62 is similar to that of well I17: by the well shut-in time, cumulative polymer injection is 203.136 t, alkaline injection is 508.593 t, and cumulative solution injection is $21.2 \times 10^4 \, m^3$. It is evaluated from the material balance method of reservoir engineering that the displacement front of combination flooding moves 101 m forward and agrees with calculation result 119 m of polymer injection front range after well shut-in of 226 h, which verified the analysis result of the testing curve.

12.2.2.2.2. Well Testing Interpretation of Well P3

- Testing survey of well P3. This well is a polymer injection well, polymer and alkaline (binary compound flooding) have been injected since May 2005. The testing period was 2006.6.22 to 2006.7.6, and testing formation was

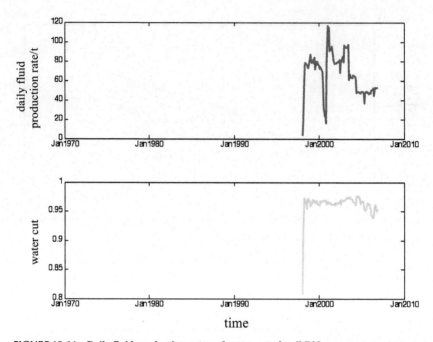

FIGURE 12.61   Daily fluid production rate and water cut of well P38.

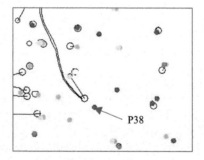

FIGURE 12.62   Well location of well P38.

upper AII layer 3 and 4. The production dynamic curve is shown in Fig. 12.69, and well location is shown in Fig. 12.70.

- Testing curve conformation analysis of well P3. In the early and middle periods, pressure and its derivation curves are two parallel straight lines showing slowly rising tendencies, which are similar to the characteristics of fracture formation. This is due to the existence of dominant flowing path in formation; at the same time, because of polymer injection, mobility deteriorates and the non-Newtonian effect appears; the pressure

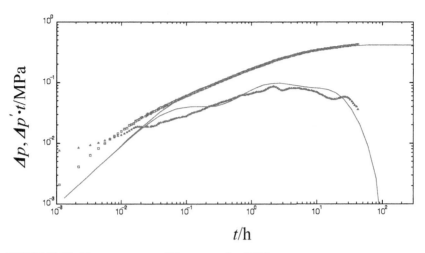

FIGURE 12.63   Pressure response fitting curve of well P38.

TABLE 12.5 Well Testing Interpretation Result of Well P38

Permeability (µm²)		Skin factor (dimensionless)		Well bore storage factor (m³ / MPa)	Investigation radius (m) (T = 2.86 h)	
All3	All4	All3	All4	0.05	All3	All4
2.44	0.90	−3.0	−3.0		185	85

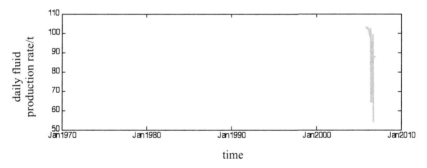

FIGURE 12.64   Production dynamic curve of I.

derivation curve drops in the late period, this variation reflects the response of polymer and oil two-phase region, and the lowest point is the polymer front. Fitting curve shape of this well (Fig. 12.71) agrees with the variation rule of the measured curve, and the interpretation result is shown in Table 12.7.

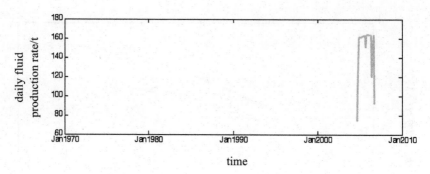

FIGURE 12.65   Production dynamic curve of I62.

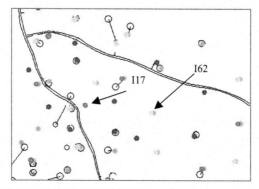

FIGURE 12.66   Well location of well I17 and I62.

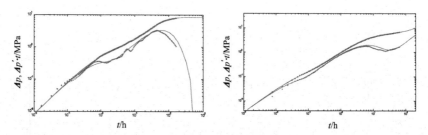

FIGURE 12.67   Pressure response fitting curve of I17.   **FIGURE 12.68   Pressure response fitting curve of I62.**

- Summary of testing results. Well P3, by the well shut-in time, cumulative polymer injection is 122.014 t, alkaline injection is 331.445 t, and cumulative solution injection is $12.7 \times 10^4 \, m^3$. It is evaluated from the material balance method of reservoir engineering that the displacement front of combination flooding moves 76 m forward and agrees with the calculation result of 84 m of the polymer injection front range after well

TABLE 12.6 Well Testing Interpretation Result of Wells I17 and I62

Well name	Parameter							
	Permeability (μm²)		Skin factor (dimensionless)		Well bore storage factor (m³/MPa)	Polymer injection front range (m)	
	AII3	AII4	AII3	AII4		AII3	AII4
I17	1.17	/	1.0	/	0.2	74	/
I62	1.58	/	−3.0	/	0.4	119	/

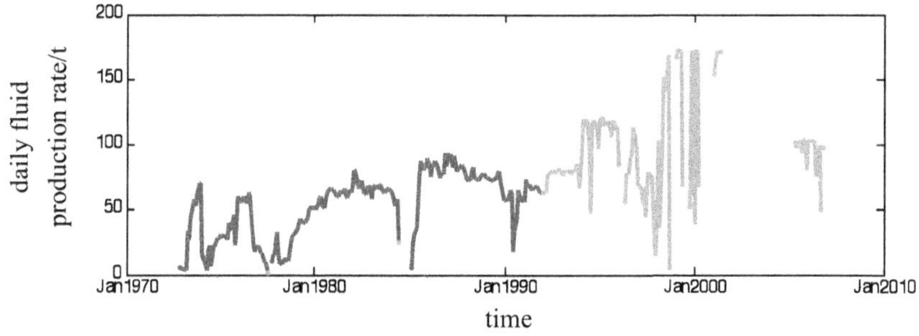

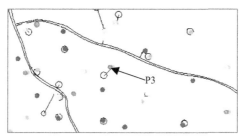

FIGURE 12.69   Production dynamic curve of well P3.

FIGURE 12.70   Well location of well P3.

shut-in of 262 h, which verified the analysis result of well testing. The pressure derivation curve drops in the late period; this is because pressure response enters the oil zone after polymer slug zone and total mobility increases, which can affect the polymer flooding front (the lowest point of pressure derivation).

12.2.2.3. Well Testing Interpretation of a Water Injection Well After Polymer Injection

- Testing survey of well P48. The testing period was 2006.7.11 to 2006.7.14; the testing interval was 1364.6–1383.8 m; the testing formation is upper AII

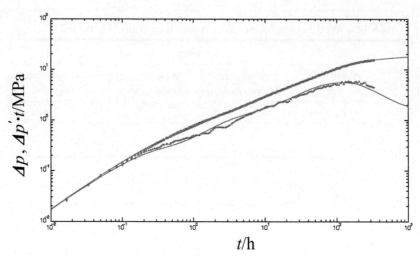

FIGURE 12.71 Pressure response fitting curve of well P3.

TABLE 12.7 Well Testing Interpretation Result of Well P3

Permeability (μm²)		Skin factor (dimensionless)		Well bore storage factor (m³ / MPa)	Polymer injection front range (m)	
AII3	AII4	AII3	AII4		AII3	AII4
1.12	1.10	0.1	0.1	0.1	98	84

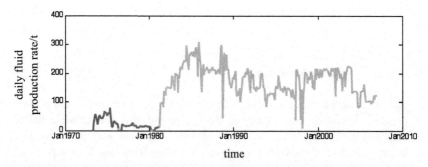

FIGURE 12.72 Production dynamic curve of well P48.

layer 3 and 4. Polymer and alkaline were injected into this well in March 1999, alkaline injection was stopped in June 2001 and polymer injection was stopped in December 2002. The production dynamic curve is shown in Fig. 12.72, and the well location is shown in Fig. 12.73.

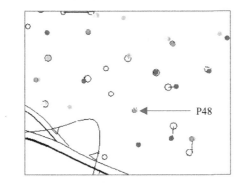

FIGURE 12.73   Well location of well P48.

- Testing Survey of Well I65. The testing period was 2006.06.06 to 06.13; testing interval was 1345.7–1383.0 m; testing formation was upper AII layer 3 and 4. Polymer and alkaline were injected into this well in March 1999, then alkaline injection was stopped in June 2001, and polymer injection was stopped in December 2002. The production dynamic curve is shown in Fig. 12.74, and the well location is shown in Fig. 12.75.
- Testing curve conformation analysis. Pressure curve conformations of well P48 and I65 are similar. The early period is influenced by the well bore storage effect, pressure and its derivation curves show rising tendencies with unity slope; the distance between pressure and its derivation curve is small and skin factor is small, which indicate a completely penetrating well or an improved well; the pressure derivation curve shows upwarp in the late period, which shows that pressure response enters the region with poor mobility after testing well shut-in, and only the pressure derivation curve of the two wells show this conformation during all testing wells. However, only the two wells injected polymer before and the polymer injection is

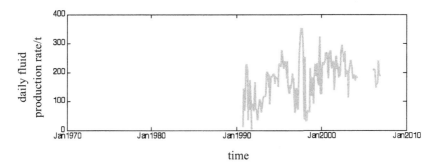

FIGURE 12.74   Production dynamic curve of well I65.

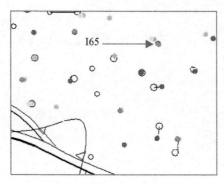

FIGURE 12.75   Well location of well I65.

stopped for a long time before testing. Because pressure derivation curve upwarp is similar to the situation where pressure response enters polymer slug, it can be inferred that there is still residual polymer in the testing well control region. The figures below show the pressure response fitting curve of the two wells, which agrees with the variation rule of the measured curve (Figs 12.76 and 12.77). The interpretation result is shown in Table 12.8.

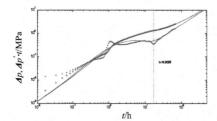

FIGURE 12.76   Pressure response fitting curve of P48.

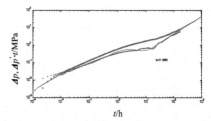

FIGURE 12.77   Pressure response curve of I65.

TABLE 12.8 Well Testing Interpretation Result of Wells P48 and I65

Well name	Parameter						
	Permeability (μm²)		Skin factor (dimensionless)		Well bore storage factor (m³ / MPa)	Polymer injection front range (m)	
	AII3	AII4	AII3	AII4		AII3	AII4
P48	0.8	0.32	−4.0	−2.0	1.0	533	260
I65	0.83	0.75	−2.5	−3.0	0.5	310	203

- Summary of testing results. The pressure derivation curves of wells P48 and I65 show upwarp in the late period, which is due to the decrease in the mobility caused by the residual polymer in the surrounding region. It can be seen from the results of the investigation radius that the residual region of polymer is 200–300 m away from the two wells.

12.2.2.4. Well Testing Interpretation of a Water Injection Well
12.2.2.4.1. Well Testing Interpretation of Well I57

- Testing survey of well I57. Well I57 was tested twice: the testing periods were 2006.7.18 to 7.22 and 2006.8.7 to 2006.8.10, respectively; the testing intervals were 1370.0–1376.7 m and 1380.6–1415.5 m, respectively; and the testing formation was upper AII layer 3 and 4. The production dynamic curve is shown in Fig. 12.78, and well location is shown in Fig. 12.79.
- Testing curve conformation analysis of well I57. The pressure response curve in the first testing period shows obvious layered testing characteristics. Pressure and derivation curves show a rising tendency with unity slope for the influence of well bore storage effect in the early

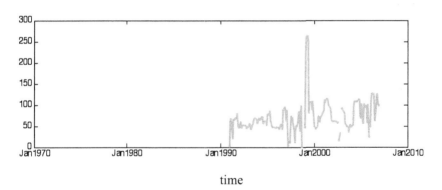

time

FIGURE 12.78   **Production dynamic curve of well I57.**

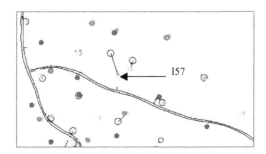

FIGURE 12.79   **Well location of well I57.**

period; the middle period is initial radial flow, only upper AII layer 3 was tested in this period, and the pressure derivation stable period corresponded to the initial radial flow period; in the later period, pressure derivation decreased indicating a transient period. In this period, because the increase of testing reservoir thickness shows vertical flow in two layers, the pressure derivation curve drops quickly; in the late period, the two layers in the double-layer reservoir both enter the radial flow stage and a stable stage appears on pressure response curve (Fig. 12.80). The interpretation result is shown in Table 12.9.

- Summary of testing results. The pressure derivation response curve of well I57 drops after testing for 31.6 h, which is influenced by the pseudo-constant pressure boundary formed by edge water, and calculation distance is 474 m. It can be seen from the well location diagram that well I57 is near eastern edge water, which agrees with the analyzed results. It can be judged that it is a pseudo-constant pressure boundary formed by edge water that causes the pressure derivation curve of well I57 to drop in the late period.

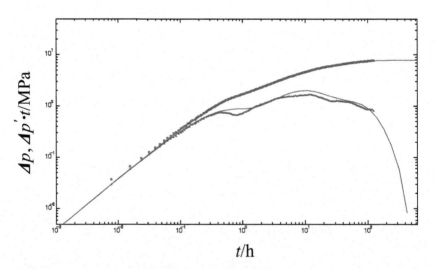

FIGURE 12.80   Pressure response fitting curve of well I57 (the first test).

TABLE 12.9 Well Testing Interpretation Result of Well I57 (the first test)

Permeability (μm²)	Skin factor (dimensionless)	Well bore storage factor (m³ / MPa)
AII3	AII3	
1.09	3.0	1.0

12.2.2.4.2. Well Testing Interpretation of Well I58

- Testing survey of well I58. Well I58 was tested twice: the testing periods were 2006.7.18 to 7.21 and 2006.8.7 to 2006.8.10, respectively; the testing intervals were 1373–1395 m and 1383.3–1395 m, respectively; and the testing formation is upper AII layer 3. The production dynamic curve is shown in Fig. 12.81, and well location is shown in Fig. 12.82.
- Testing curve conformation analysis of well I58. The two measured pressure curve conformations of well I58 are similar: with the influence of well bore storage effect in the early period, pressure and its derivation curve show a rising tendency with unity slope; and it enters the late period after a short middle radial flow period (Fig. 12.83); in the late period, the pressure derivation curve shows an obvious rising tendency with the influence of the western fault. In the two measured pressure response curves, the opening of the pressure and derivation curve is very large, which indicates that well I58 is an imperfect well, and there is pollution around the well bore. Fitting curve conformation of this well agrees with variation rule of the measured curve, the interpretation result is shown in Table 12.10.
- Summary of testing results. The two well testing interpretation results of well I58 are similar: in the late period, the calculation results of the investigation radius are 86 m and 84 m, respectively. This agrees with the western fault

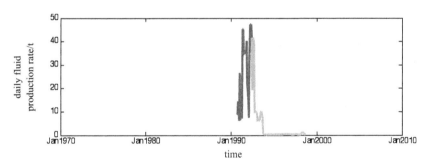

FIGURE 12.81   Production dynamic curve of well I58.

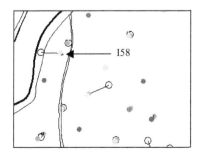

FIGURE 12.82   Well location of well I58.

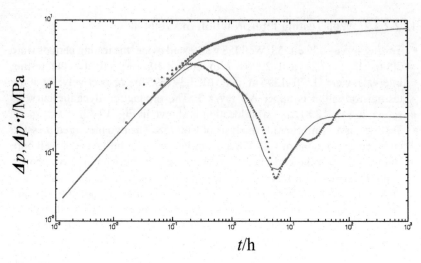

FIGURE 12.83   Pressure response fitting curve of well I58 (the first test).

TABLE 12.10 Well Testing Interpretation Result of Well I58 (the first test)

Permeability (μm²)		Skin factor (dimensionless)		Well bore storage factor (m³/MPa)	Polymer injection front range (m)	
AII3	AII4	AII3	AII4		AII3	AII4
0.71	/	3.0	/	0.1	86	/

89 m of well I58 in the seismic data. Hence, the calculated fault range accords with the geological structure.

12.2.2.4.3. Well Testing Interpretation of Well P67

- Testing of well P67. The testing period was 2006.7.111 to 7.22; the testing interval was 1365.6–1385 m; testing formation was upper layer AII3 and 4. The production dynamic and well location are shown in Figs 12.84 and 12.85, respectively.
- Testing curve conformation analysis of well P67. The pressure derivation curve of this well varies a lot in the middle period. There are two main reasons for this: the first is the complexity of formation properties, small faults with poor closeness exists around; the second is the influence of pseudo-constant pressure boundary formed by other water injection wells and northeast edge water (Fig. 12.86). A well testing interpretation model is difficult to describe these two reasons quantitatively, so the pressure derivation curve is difficult to fit. The second upwarp of the pressure

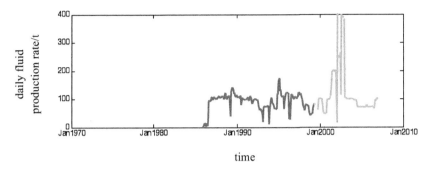

FIGURE 12.84   Production dynamic curve of well P67.

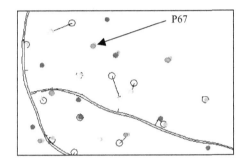

FIGURE 12.85   Well location of well P67.

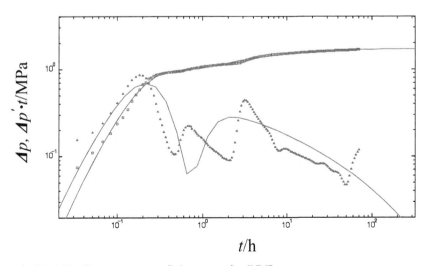

FIGURE 12.86   Pressure response fitting curve of well P67.

derivation curve in the late period is influenced by the western secondary fault of the testing well. Fitting curve conformation of this well agrees with the variation rule of the measured curve basically and interpretation results are shown in Table 12.11.

- Summary of testing results. The testing curve of well P67 is complex in the middle period. After 2.14 h of well shut-in, the pressure derivation curve shows a sharp upwarp. It can be judged from the calculation result of the investigation radius that: small faults with poor closeness exist 70 m around the testing well. Meanwhile, after well shut-in testing for 3.1 h, the pressure derivation curve drops sharply, and the calculation result of the investigation radius is 86 m. Analysis shows that the reason for the drop in the pressure derivation curve in the testing well is that well P67 and its southwest 170 m well P73 are injection wells, and no production well exists between two injection wells; this causes the energy yielding between the injection wells and forms a pseudo-constant pressure boundary; after 50.1 h of well shut-in, the pressure derivation curve shows an upwarp again, which is caused by the influence of the western secondary fault of the testing well and the calculated fault range of 353 m agrees with the measured 369 m by seismic data.

12.2.2.4.4. Well Testing Interpretation of Wells P65 and I67

- Testing of well P65 and I67. The testing periods were 2006.7.7 to 2006.7.10 and 2006.7.24 to 2006.7.27, respectively; the testing intervals were 1359.4–1365 m and 1349.1–1365.2 m, respectively; the testing formation was upper layer AII3 and 4. The production dynamic curves are shown in Figs 12.87 and 12.88, respectively, and well location is shown in Fig. 12.89.
- Testing curve conformation analysis of well P65 and I67. Wells P65 and I67 are located in the north of the reservoir; the distance between the two wells is not very far (about 300 m); and the formation characteristics are similar; so the pressure response curve conformation is similar too—there is no radial flow straight segment after the early period, and the pressure derivation curve rises after the sharp drop (Figs 12.90 and 12.91). Analysis shows that because wells P65 and I67 are separate injection wells, after a certain period of well shut-in, inter-layer cross-flow (which is caused by imbalance of inter-layer

TABLE 12.11 Well Testing Interpretation Result of Well P67

Permeability (μm^2)		Skin factor (dimensionless)		Well bore storage factor (m^3 / MPa)	Fault range (m)		Constant pressure boundary range (m)		Fault range (m)	
AII3	AII4	AII3	AII4		AII3	AII4	AII3	AII4	AII3	AII4
0.5	0.1	1.5	2.5	0.15	181	71	219	86	890	353

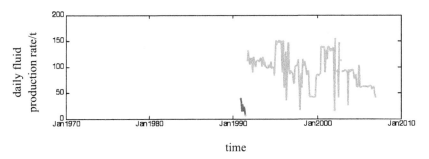

FIGURE 12.87 **Production dynamic curve of well P65.**

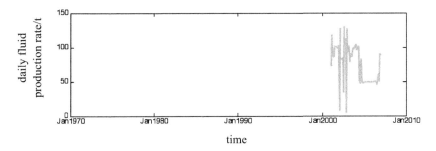

FIGURE 12.88 **Production dynamic curve of well I67.**

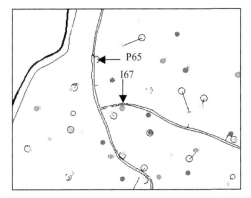

FIGURE 12.89 **Well location of P65 and I67.**

pressure) is the main reason for the upwarp in the pressure derivation curve. In the late testing period, the pressure derivation curve of well P65 shows an obvious rising tendency, which is different from the slow tendency of the pressure response curve of well I67. Analysis shows that this is influenced by the western main fault of well P65 and causes an upwarp in the pressure derivation curve. Interpretation results are shown in Table 12.12.

- Summary of testing results. The corresponding investigation radius calculated result of the well P65 upwarp point is 274 m and it can be seen

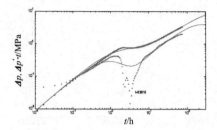

FIGURE 12.90 Pressure response fitting curve of P65.

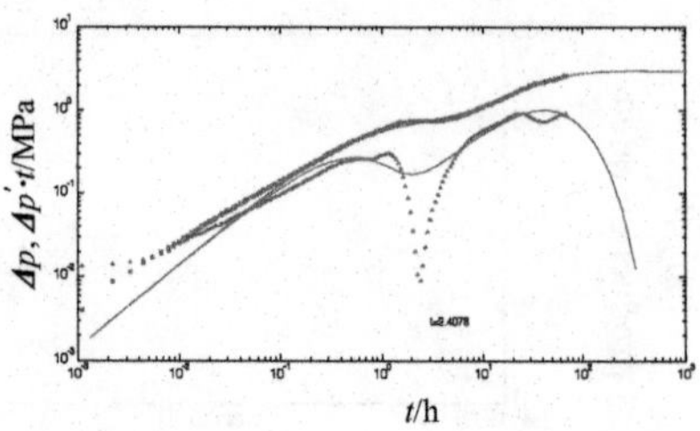

FIGURE 12.91 Pressure response fitting curve of I67.

TABLE 12.12 Well Testing Interpretation Result of Wells P65 and I67

Well name	Parameter						
	Permeability (μm²)		Skin factor (dimensionless)		Well bore storage factor (m³ / MPa)	Polymer injection front range (m)	
	AII3	AII4	AII3	AII4		AII3	AII4
P65	1.13	1.09	−3.0	−4.0	0.6	280	274
I67	0.40	0.90	−4.0	−4.0	0.1	/	/

from the sand microstructure diagram that the fault is located in the west of well P65 250 m. The calculated fault range agrees with the structure situation. Then testing curve conformation analysis result is verified: the pressure derivation curve upwarp of well P65 in the late period is influenced by the western fault.

12.2.2.4.5. Well Testing Interpretation of Well I43

- Testing of well I43. The testing period was 2006.7.18 to 2006.7.21; the testing interval was 1346.2–1381.2 m; the testing formation was upper AII 3 and 4. The production dynamic curve is shown in Fig. 12.92, and well location is shown in Fig. 12.93.
- Testing curve conformation analysis of well I43. The pressure response curve of this well shows a slowly rising tendency in the early period and the enclosed area of the pressure and derivation curve is small, which indicates that this well condition is good and that there is no pollution. The pressure derivation curve shows an obvious upwarp in the middle period; it can be judged that this is influenced by the western main fault of well I43 in the late period and the pressure derivation curve shows an upwarp again with the influence of the eastern secondary fault (Fig. 12.94). Fitting curve conformation of this well basically accords with the variation rule of the measured curve and the interpretation results are shown in Table 12.13.

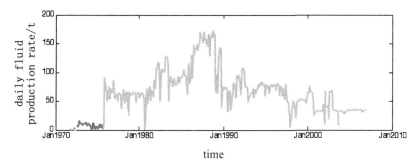

FIGURE 12.92   Production dynamic curve of well I43.

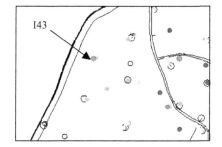

FIGURE 12.93   Well location of well I43.

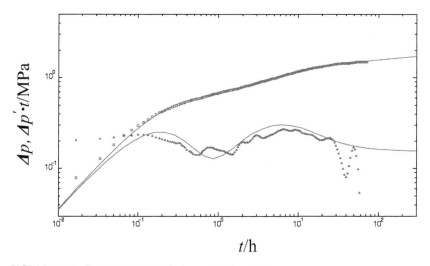

FIGURE 12.94   Pressure response fitting curve of well I43.

TABLE 12.13 Well Testing Interpretation Result of Well I43

Permeability (μm^2)		Skin factor (dimensionless)		Well bore storage factor (m^3 / MPa)	Fault range (m)	
AII3	AII4	AII3	AII4		AII3	AII4
1.05	1.17	−3.0	−2.0	0. 4	95	103

12.2.2.4.6. Well Testing Interpretation of Well I64

- Testing of well I64. The testing period was 2006.8.7 to 2006.8.10; the testing interval was 1371.3–1399.4 m; the testing formation was upper AII3; the production dynamic curve is shown in Fig. 12.95, and well location is shown in Fig. 12.96.
- Testing curve conformation analysis of well I64. The pressure and derivation curves of well I64 are soft, there is a small hump in the pressure derivation curve, which indicates that there is light pollution around the well bore (Fig. 12.97). The fitting curve conformation of this well accords with the

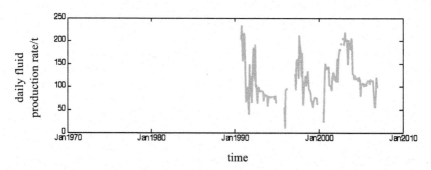

FIGURE 12.95   Production dynamic curve of well I64.

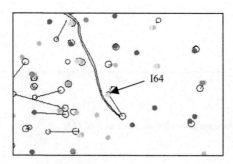

FIGURE 12.96   Well location of well I64.

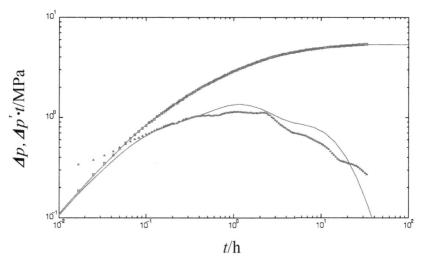

FIGURE 12.97   Pressure response fitting curve of well I64.

variation rule of measured curve and the interpretation result is shown in Table 12.14.

12.2.2.4.7. Well Testing Interpretation of Wells P80 and P56

- Testing of wells P80 and I56. The testing periods were 2006.7.7 to 7.10 and 2006.7.22–2006.7.26, respectively; the testing intervals were 1381.35–1386.15 m and 13711–1394 m, respectively; the testing formation of well P80 was upper AII3 and the testing formation of well P56 was upper AII4; the production dynamic curves are shown in Figs 12.98 and 12.100, respectively, and well location is shown in Fig. 12.99 and 12.101, respectively.
- Pressure response curve conformation analysis of well P80 and P56. Well P80 is located around the northeast edge water, well P56 is located around the southeast edge water and their pressure response curve conformations are similar: the pressure derivation curve drops in the late period, which is the influence of the pseudo-constant pressure boundary formed by edge water; the pressure derivation drop degree of well P80 is greater than that of well

TABLE 12.14 Well Testing Interpretation Result of Well I64

Permeability (μm²)		Skin factor (dimensionless)		Well bore storage factor (m³ / MPa)
AII3	AII4	AII3	AII4	
0.82	/	0.5	/	0.2

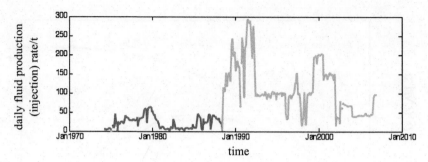

FIGURE 12.98   Production dynamic curve of well P80.

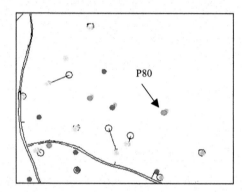

FIGURE 12.99   Well location of well P80.

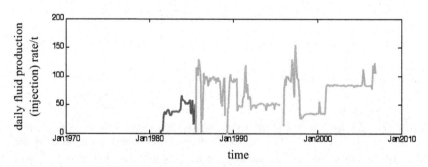

FIGURE 12.100   Production dynamic curve of well P56.

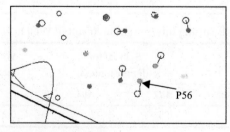

FIGURE 12.101   Well location of well P56.

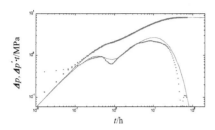

FIGURE 12.102 Pressure response fitting curve of P80.

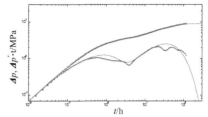

FIGURE 12.103 Pressure response fitting curve of P56.

P56 and it indicates that the northeast edge water around well P80 is greater than the southwest edge water around well P56 (Figs 12.102 to 12.103). Fitting curve conformation of this well basically accords with the variation rule of the measured curve and the interpretation result is shown in Table 12.15.

- Summary of testing result. The pressure derivation curve of well P80 drops after well shut-in of 22.5 h, which is the influence of the constant boundary formed by edge water. The calculated range is 198 m, that can be seen from the well location diagram and well P80 is located near the eastern edge water, which accords with the analysis results.
- The pressure derivation curve of well P56 drops in the late period, which is the influence of the constant boundary formed by the edge water; the calculated range is 209 m, which can be seen from the well location diagram and well P80 is located near the southeast edge water, which accords with the analysis results.

12.2.2.4.8. Well Testing Interpretation of Well P73

- Testing of well P73. The testing period was 2006.7.11 to 7.14; the testing interval was 1356.6–1372.6 m; testing formation is upper AII3; the

TABLE 12.15 Well Testing Interpretation Result of Wells P80 and P56

Well name	Parameter								
	Permeability (μm^2)		Skin factor (dimensionless)		Well bore storage factor (m^3 / MPa)	Fault range (m)		Constant pressure boundary range (m)	
	AII3	AII4	AII3	AII4		AII3	AII4	AII3	AII4
P80	0.64	/	−2.3	/	2.0	301	/	198	/
P56	0.26	0.2	−3.0	−1.0	2.5	/	/	209	230

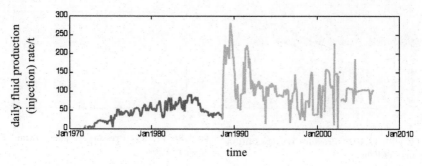

FIGURE 12.104   Production dynamic curve of well P73.

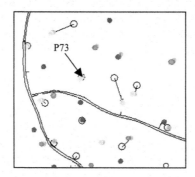

FIGURE 12.105   Well location of well P73.

production dynamic curve is shown in Fig. 12.104, and well location is shown in Fig. 12.105.

- The pressure response curve conformation analysis of well P73. The pressure and derivation curves show rising tendencies as two parallel straight lines in the early and middle periods, which is because the existence of the dominant flowing path makes the pressure wave transmit faster than in other regions, and pressure falls quickly; the pressure derivation curve falls in the late period, which is the influence of the pseudo-constant pressure formed by the surrounding water injection wells (Fig. 12.106). The fitting curve conformation of this well basically accords with the variation rule of the measured curve and the interpretation result is shown in Table 12.16.

- Summary of testing result. After well P73 shut-in of 31.6 h, the pressure derivation curve begins to fall and the calculated investigation radius is 118 m. The reason for the drop in the testing well pressure derivation curve is that well P73, and its northeast 190 m well P67, are water injection wells, and there is no production well between two water injection wells. Energy balance is formed between two injection wells and a pseudo-constant pressure boundary is formed too. The calculation result basically accords with the well location situation.

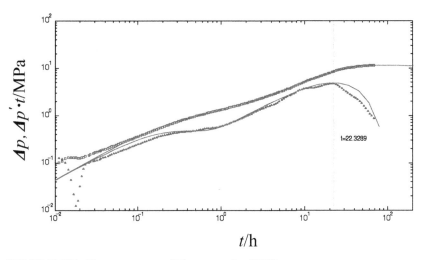

FIGURE 12.106   Pressure response fitting curve of well P73.

TABLE 12.16   Well Testing Interpretation Result of Well P73

Permeability (μm²)		Skin factor (dimensionless)		Well bore storage factor (m³/MPa)	constant pressure boundary range/m	
AII3	AII4	AII3	AII4		AII3	AII4
1.12	/	5.0	/	2.5	118	/

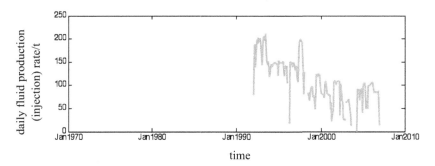

FIGURE 12.107   Production dynamic curve of well I27.

12.2.2.4.9. Well Testing Interpretation of Well I27

- Testing of well I27. The testing period was 2006.11.14 to 2006.11.28; the testing interval was 1346.6–1360.3 m; the testing formation is upper AII3 and 4. The production dynamic curve is shown in Fig. 12.107, and well location is shown in Fig. 12.108.

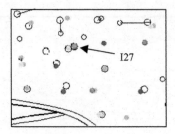

FIGURE 12.108   Well location of well I27.

- Pressure response curve conformation analysis of well I27. In the early period, with the influence of well bore storage effect, the pressure and derivation curve shows a rising tendency with a small slope, which indicates that the permeability is good around this well and oil well blowdown is fast. In the middle period, the pressure derivation curve falls (Fig. 12.109). This is because well I27 is near the western gas region and the channel gas of the testing well around the formation causes the rise in total mobility and the decline in the pressure derivation curve. In the late period, the pressure derivation curve shows an upwarp after radial flow, but the range is not very large, which shows that this is the interference influence of continuous production in the well shut-in period. The fitting curve conformation of this well basically accords with the variation rule of the measured curve and the interpretation result is shown in Table 12.17.

12.2.2.4.10. Well Testing Interpretation of Well P28

- Testing of well P28. The testing period was 2006.7.12 to 7.14; the testing interval was 1359.6–1264.6 m; the testing formation of well P28 was upper

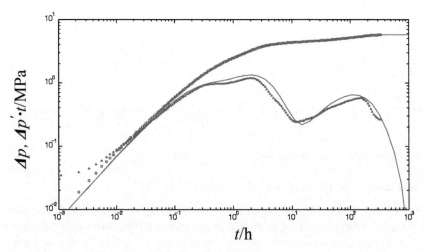

FIGURE 12.109   Pressure response fitting curve of well I27.

TABLE 12.17 Well Testing Interpretation Result of Well I27

Permeability (μm^2)		Skin factor (dimensionless)		Well bore storage factor (m^3 / MPa)
AII3	AII4	AII3	AII4	
1.01	0.08	2.0	2.0	0.26

AII3 and 4. The production dynamic curve is shown in Fig. 12.110, and the well location is shown in Fig. 12.111.

- Pressure response curve conformation analysis of well P28. The pressure and derivation curves show rising tendencies as two parallel straight lines in the early and middle periods (Fig. 12.112). This is because the existence of the dominant flowing path makes the pressure wave transmit faster than in other regions, and pressure falls faster; the pressure derivation curve falls in the late period. This is the interference influence of continuous production in well

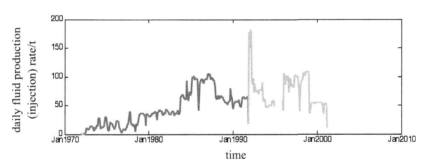

FIGURE 12.110   Production dynamic curve of well P28.

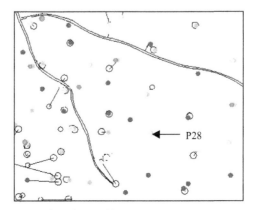

FIGURE 12.111   Well location of well P28.

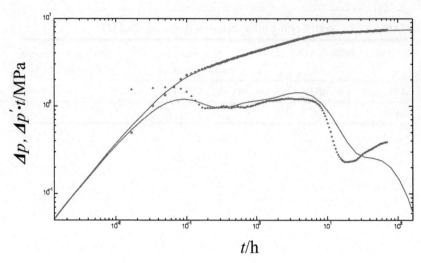

FIGURE 12.112 Pressure response fitting curve of well P28.

shut-in period. The fitting curve conformation of this well basically accords with the variation rule of the measured curve and the interpretation result is shown in Table 12.18.

12.2.2.5. Well Testing Interpretation of the Interference Testing Well Group

Take well P11 as the active well and wells I17, P3, P15, I62 as observation wells, then interference testing starts and testing procedures are shown below:

A high-precision electronic gauge is laid into active well P11 and observation wells I17, P3, P15, P16. Shut-in all wells after 4 h. Keep the working system of the surrounding well unchanged and observe for three days for pressure tendency, then take out a pressure gauge to observe. Wait until stable reservoir pressure tendency is reached, shut-in well P11 and take interference testing of well P11 and the four observation wells. Keep the working system of the surrounding well unchanged in the interference testing period. After a certain period of well shut-in, take one pressure gauge from an observation well and

TABLE 12.18 Well Testing Interpretation Result of Well P28

Permeability (μm²)		Skin factor (dimensionless)		Well bore storage factor (m³ / MPa)
AII3	AII4	AII3	AII4	
0.52	3.07	2.2	3.0	0.05

observe the pressure curve: if it achieves the requirements of well testing interpretation, take the pressure gauges of the other wells and observe the pressure curves. If the testing pressure of each well has achieved the requirements of well testing interpretation, recover the production; if they have not achieved the requirements of interpretation, lay the pressure gauge again and keep measuring until all of them achieve the requirement.

12.2.2.5.1. Well Testing Interpretation of Active Well P11

- Testing of well P11. Oil well P11 was taken as the active well of this interference testing. The testing period was 2006.6.21 to 2006.6.28; the testing formation is upper AII 3 and 4. The production dynamic curve is shown in Fig. 12.113 and well location is shown in Fig. 12.114.
- Pressure response curve conformation analysis of well P11. The pressure and derivation curves of well P11 change softly, and the pressure response enters the radial flow period very late; pressure derivation varies in a complex way in the radial flow period and this is caused by strong heterogeneity distribution of fluid in polymer injection region (wells I17 and I62) around well P11 (Fig. 12.115). The pressure derivation curve falls in the late period; this is due to the second pulse. The fitting curve conformation of this well basically accords with the variation rule of the measured curve and the interpretation result is shown in Table 12.19.

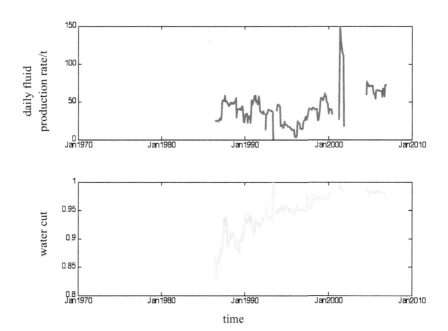

FIGURE 12.113   Production dynamic curve of well P11.

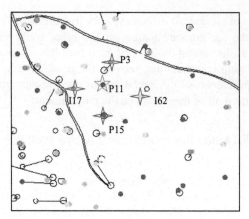

FIGURE 12.114   Well location of well P11.

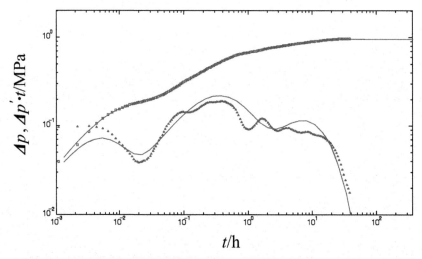

FIGURE 12.115   Pressure response fitting curve of well P11.

TABLE 12.19 Well Testing Interpretation Result of Well P11

Permeability (μm²)		Skin factor (dimensionless)		Well bore storage factor (m³ / MPa)
AII3	AII4	AII3	AII4	
0.27	1.05	−3.0	−3.0	0.1

12.2.2.5.2. Well Testing Interpretation of Observation Well P15

- Survey of well P15. Oil well P15 was taken as the observation well of the interference testing; the testing period was 2006.6.20 to 2006.7.6; the testing formation was upper AII3 and 4. The production dynamic curve is shown in Fig. 12.116 and well location is shown in Fig. 12.117.

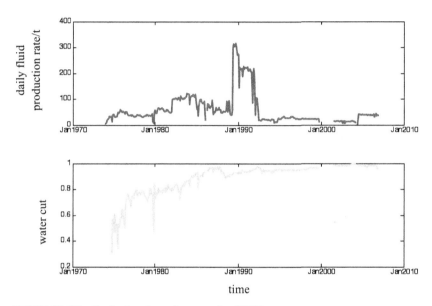

FIGURE 12.116   Production dynamic curve of well P15.

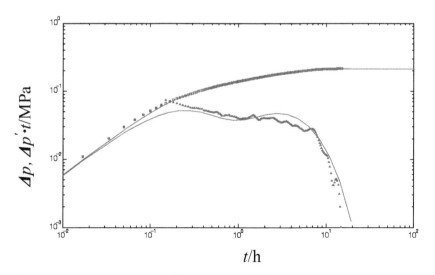

FIGURE 12.117   Pressure response fitting curve of well P15.

- Pressure response curve conformation analysis of well P15. The pressure and derivation curves of well P15 change softly; the pressure derivation curve falls before entering the radial flow period. This is because the surrounding injection wells shut-in during the process of the interference test, and the formation pressure drops totally. The Fitting curve conformation of this well basically accords with the variation rule of the measured curve and the interpretation result is shown in Table 12.20.

2.2.5.3. Well Testing Interpretation of Observation Wells I17 and I62

- Survey of well I17 and I62. Wells I17 and I62 are injection wells. For well I17, polymer and alkaline were injected simultaneously (binary compound flooding); for well I62, polymer was injected in May 2005 and alkaline was injected in January 2006 (binary compound flooding). Both the testing periods of the two wells were 2006.6.22 to 2006.7.6; the testing formation of well I17 was upper AII3; the testing formation of well I62 was upper layer AII3 and 4. Production dynamic curves are shown in Figs 12.118 and 12.119, and well location is shown in Fig. 12.114.
- Testing curve conformation analysis of well I17 and I62. The pressure response curve conformation of wells I17 and I62 are similar. The pressure and derivation curves show slowly rising tendencies as two parallel straight lines, which are similar to fracture formation characteristics. Analysis shows that the reason for this curve characteristic is the existence of the dominant flowing path in formation. Meanwhile, mobility becomes worse and

TABLE 12.20 Well Testing Interpretation Result of Well P15

Permeability (μm^2)		Skin factor (dimensionless)		Well bore storage factor (m^3 / MPa)
AII3	AII4	AII3	AII4	
0.70	0.83	−3.0	−3.0	0.2

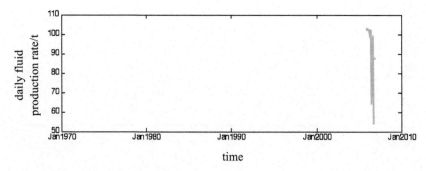

FIGURE 12.118 Production dynamic curve of well I17.

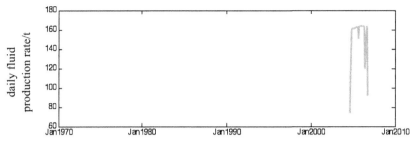

FIGURE 12.119   Production dynamic curve of well I62.

causes non-Newtonian fluid effects with polymer injection. The pressure derivation curve falls in the late period; analysis shows that this is caused by the increase in total mobility while the pressure response enters the oil zone after the polymer zone (Figs 12.120 to 12.121). The lowest point is the polymer front; the pressure derivation curve shows upwarp at the end. This is caused by the second pulse interference in the late period when active well P11 begins to produce. The fitting curve conformation basically accords with the variation rule of the measured curve; the interpretation result is shown in Table 12.6.

12.2.2.5.4. Well Testing Interpretation of Observation Wells P3

- Testing of well P3. This well is a polymer injection well: polymer and alkaline were injected in May 2005 (binary compound flooding). The testing period was 22 June 2006 to 6 July 2006; the testing interval was upper AII3 and 4. The production dynamic curve is shown in Fig. 12.122 and well location is shown in Fig. 12.114.
- Testing curve conformation analysis of well P3. The pressure and derivation curves show slowly rising tendencies as two parallel straight lines in the early and middle periods, which is similar to fracture formation characteristics. Analysis shows that the cause of this curve characteristic is the existence of the dominant flowing path in formation.

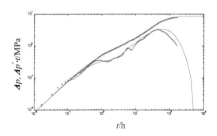

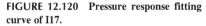

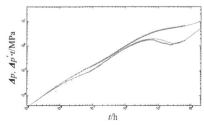

FIGURE 12.120   Pressure response fitting curve of I17.

FIGURE 12.121   Pressure response fitting curve of I62.

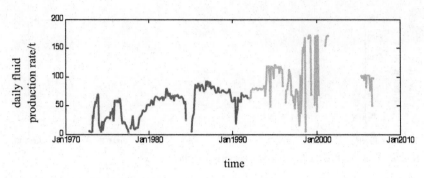

FIGURE 12.122   Production dynamic curve of well P3.

Meanwhile, mobility deteriorates and causes non-Newtonian fluid effects with polymer injection. The pressure derivation curve falls in the late period; analysis shows that this is caused by polymer and oil two-phase region response (Fig. 12.123). The fitting curve conformation basically accords with the variation rule of the measured curve; the interpretation result is shown in Table 12.21.

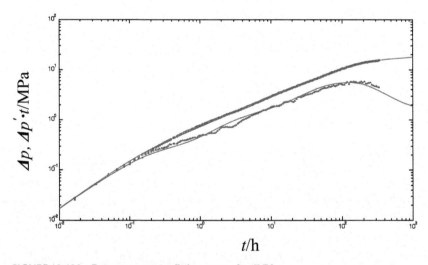

FIGURE 12.123   Pressure response fitting curve of well P3.

TABLE 12.21 Well Testing Interpretation Result of Well P3

Permeability (μm²)		Skin factor (dimensionless)		Well bore storage factor (m³/MPa)
AII3	AII4	AII3	AII4	
1.12	1.10	0.1	0.1	0.1

TABLE 12.22 Well Testing Interpretation Results

No.	Well number	Interpretation formation	Permeability (μm^2)	Skin factor (dimensionless)	Well bore storage factor (m^3 / MPa)	Investigation radius (m)	
1	P67	AII3 and 4	0.56&0.11	1.5&2.5	0.15	71	86
2	P80	AII3	0.64	−2.3	2.0	187	
3	P73	AII3	1.12	5.0	2.5	218	
4	I57	AII3	1.09	12.5	1.0	474	
5	I57	AII4	1.28	3.0	0.8	/	
6	I27	AII3 and 4	1.01&0.08	2.0&2.0	0.26	/	
7	P85	AII3	1.43	−3.0	0.25	121	
8	P28	AII3 and 4	0.52&3.07	2.2&3.0	0.05	323	
9	P38	AII3 and 4	2.44&0.9	−3.0&−3.0	0.05	93	
10	I65	AII3 and 4	0.83&0.75	−2.5&−3.0	0.5	203	
11	I43	AII3 and 4	1.05&1.17	−3.0&−2.0	0.4	95	
12	I58	AII3	0.71	3.0	0.1	86	
13	I58	AII3	0.70	5.0	0.05	84	
14	P12	AII3 and 4	0.62&0.6	−0.5&0	0.42	/	
15	P65	AII3 and 4	1.13&1.09	−3.0&−4.0	0.6	274	
16	I67	AII3 and 4	0.4&0.9	−4.0&−4.0	0.1	/	
17	P56	AII3 and 4	0.26&0.2	−3.0&−1.0	2.5	209	
18	P81	AII3 and 4	1.46&0.87	−3.0&−2.0	0.4	132	
19	P48	AII3 and 4	0.8&0.32	−4.0&−2.0	1.0	52	260
20	P35	AII3 and 4	0.68&1.44	−1.0&−2.0	0.4	147	
21	I64	AII3	0.82	0.5	0.2	/	
22	P15	AII3 and 4	0.7&0.83	−3.0&−3.0	0.2	239	
23	P11	AII3 and 4	0.27&1.05	−3.0&−3.0	0.1	131	
24	I62	AII3	1.58	−3.0	0.4	119	
25	P3	AII3 and 4	1.12&1.1	0.1&0.1	0.1	/	
26	I17	AII3	1.17	1.0	0.2	74	

2.2.6. Comprehensive Interpretation of Total Wells

Well testing interpretation results of all testing wells is shown in Table 12.22; permeability distribution after interpretation is shown in Figs 12.124 and 12.125, respectively.

The last production period of this study is in December 2006. By this time, the remaining geological reserves of upper AII layer 3 and 4 are 393.33×10^4 t and 115.45×10^4 t, respectively; the total remaining geological reserves are 508.78×10^4 t; the recovery degrees of upper AII layers 3 and 4 are 45.86% and 40.46%, respectively, and total recovery degree is 44.72%.

It can be seen from the oil saturation distribution shown in Figs 12.126 and 12.127, that the remaining oil-enriched areas of upper AII layer 3 are mainly concentrated in several regions as below, by December 2006: (1) reservoir side, because of fault block, injection water is difficult to achieve and this is the

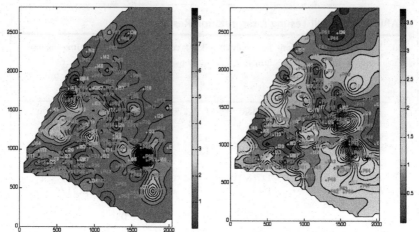

FIGURE 12.124 Permeability distribution of comprehensive interpretation AII3 (μm^2).

FIGURE 12.125 Permeability distribution of comprehensive interpretation AII4 (μm^2).

remaining oil-enriched area; (2) the flooding pattern in the western region of well I24 is imperfect, water flooding can not sweep or only a small region is swept; there is no streamline or just a few streamlines, which is also the remaining oil-enriched area; (3) flooding pattern in the northeast region of well P68 is imperfect, rare streamline and low displacement efficiency forms the remaining oil-enriched area; (4) from well P85–P17–P29 to the surrounding region of well P54 in the reservoir southeast, this region is near the reservoir boundary with high production and low injection; the well pattern is imperfect, and there is a place that streamline can not sweep, then it is an important remaining oil-enriched area.

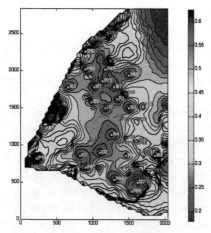

FIGURE 12.126 Oil saturation distribution of comprehensive interpretation AII3.

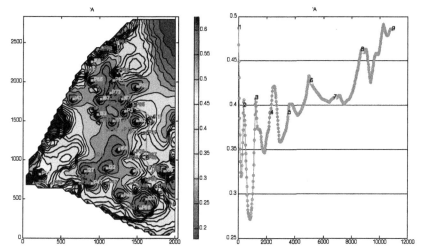

FIGURE 12.127   Oil saturation profile of comprehensive interpretation AII3.

The remaining oil-enriched areas of upper AII layer 4 are mainly concentrated in several regions as below: (1) reservoir side, because of fault block, injection water is difficult to achieve and this is the remaining oil-enriched area; (2) enclosed area of well I43–P81–P35–P12, no injection and production well, oil producing degree is low and it is the main remaining oil-enriched area; (3) the eastern region of well P39–P51, no injection, and near the reservoir boundary streamline is sparse, which is an important remaining oil-enriched area (see Figs 12.128 to 12.131).

12.3. APPLICATION CASE THREE

12.3.1. Survey of Block Geology and Development

The study block of application 3 is a natural block cut by fault, which is a high-porosity and high-permeability reservoir. Meandering river and braid river are the main sediment characteristics, sandy body is scattered developed, plane heterogeneity is serious, reservoir burial depth is shallow, cementation is loose, mud cement is high, and sand production is serious. East, south and north are all faults. The characteristics of this block are shallow reservoir burial depth, bad compaction and easy sand production. The oil area is 7.0 km^2, effective thickness is 18.4 m, geological reserve is 2541×10^4 t, and available reserve is 610×10^4 t. Crude oil is composed of low-freezing heavy aromatics with low sulfur and paraffin content, underground oil density is $0.8939 \, \text{g} / \text{cm}^3$, underground oil viscosity is 75.5 mPa • s, and underground oil/water viscosity ratio is 171.6. Surface oil viscosity distribution differs greatly and varies with structure height, variation range is 550–2480 mPa • s, oil property in the southwest is heavy and the viscosity is above 2000 mPa • s. Initial reservoir pressure is

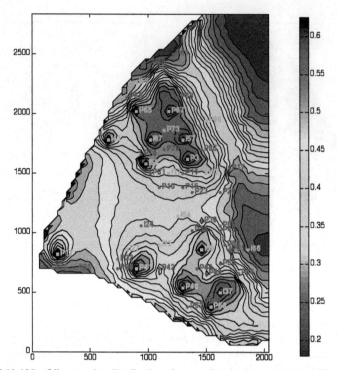

FIGURE 12.128   Oil saturation distribution of comprehensive interpretation AII4.

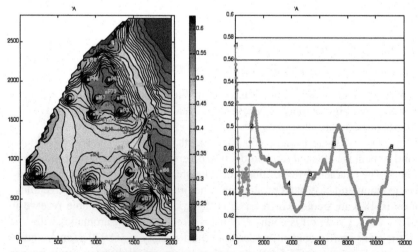

FIGURE 12.129   Oil saturation profile of comprehensive interpretation AII4.

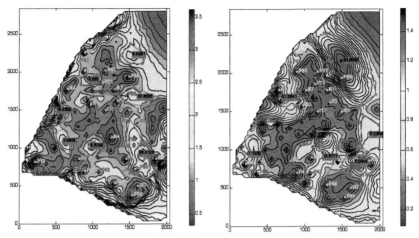

FIGURE 12.130 Reserves abundance distribution of comprehensive interpretation AII3 $(t \bullet m^{-2})$.

FIGURE 12.131 Reserves abundance distribution of comprehensive interpretation AII4 $(t \bullet m^{-2})$.

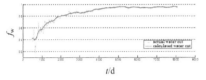

FIGURE 12.132 Block composite water-cut fitting curve.

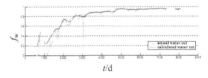

FIGURE 12.133 Water-cut fitting curve of P10.

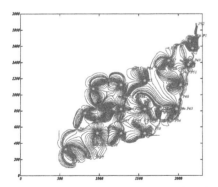

FIGURE 12.134 Streamline distribution of AII5 at the testing period.

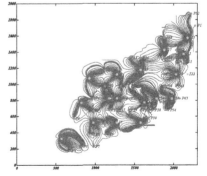

FIGURE 12.135 Streamline distribution of AII6 at the testing period.

13.0 MPa, saturation pressure is 10.5–12 MPa, so the difference between reservoir pressure and saturation pressure is 1.0–2.5 MPa.

This block has been developing since 1986, and there are four development periods including producing test and deliverability construction, water injection response, pattern modification, and comprehensive modification. Then full scale polymer flooding oil production began in December 2006. At the beginning of the production of this block, two layer systems were used for development including upper AII layer 3, 4 and upper AII layer 5, 6, the upper AII layer 5, 6 system is the research objective, there are 70 production wells and 37 injection wells in the history of this layer system.

In recent years, because of the low re-perforating potential and bad measure effect, production declines very quickly, and now it is under the extra-high water cut stage. Through measures including renewing of accident well, changing oil well to injection well, and migration, the polymer injection scheme is improved,

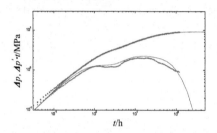

FIGURE 12.136 Pressure difference and derivation fitting curve of I7.

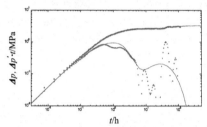

FIGURE 12.137 Pressure difference and derivation fitting curve of I13.

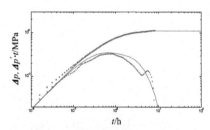

FIGURE 12.138 Pressure difference and derivation fitting curve of I14.

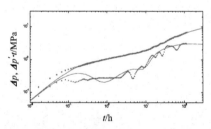

FIGURE 12.139 Pressure difference and derivation fitting curve of I16.

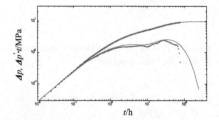

FIGURE 12.140 Pressure difference and derivation fitting curve of I18.

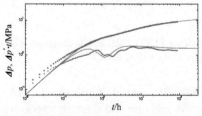

FIGURE 12.141 Pressure difference and derivation fitting curve of I19.

which successfully insures polymer injection. To study the residual oil distribution after polymer flooding, 10 wells are chosen for pressure build-up and pressure drawdown well testing.

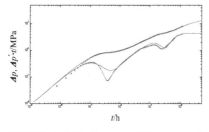

FIGURE 12.142 Pressure difference and derivation fitting curve of I23.

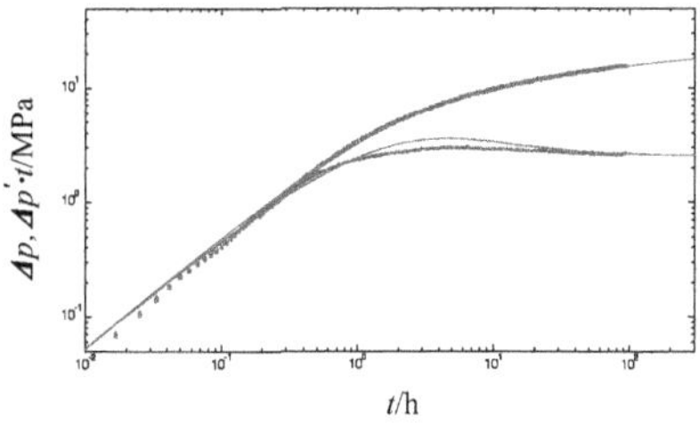

FIGURE 12.143 Pressure difference and derivation fitting curve of P3.

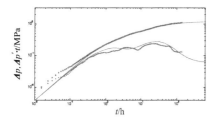

FIGURE 12.144 Pressure difference and derivation fitting curve of P17.

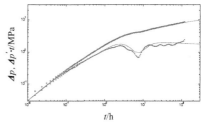

FIGURE 12.145 Pressure difference and derivation fitting curve of P29.

TABLE 12.23 Interpretation Result of Testing Well in Application Case 3

Well name	Parameter				
	Well point permeability (μm^2)		Damage factor (dimensionless)		Well bore storage factor (m^3/MPa)
	AII5	AII6	AII5	AII6	
I7	0.87	1.21	−2.02	−1.31	0.408
I13	0.42	3.32	0.14	0.55	0.235
I14	0.44	2.82	−3.61	−3.33	0.466
I16	0.47	6.02	−5.23	−3.29	0.247
I18	0.45	0.82	−2.97	0.27	0.322
I19	3.02	0.93	−1.38	−3.06	0.276
I23	1.87	3.17	−4.38	−3.68	0.071
P3	6.06	3.96	−1.55	0.17	0.148
P17	0.79	1.47	−4.25	−5.27	0.654
P29	0.73	0.46	0.04	−1.23	0.112

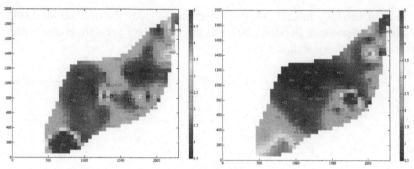

FIGURE 12.146 Permeability distribution of AII5 at the testing period (μm²).

FIGURE 12.147 Permeability distribution of AII6 at the testing period (μm²).

12.3.2. Streamline Numerical Well Testing Interpretation

The model is a two-layer reservoir and a uniform grid system is used in the plane: there are 46 grids in direction X with grid steps of 50 m; there are 40 grids in direction Y with grid steps of 50 m; there are two grids in direction Z, they are not uniform grids with effective thickness as grid step. The total grids node is $46 \times 40 \times 2 = 3680$. Kriging interpolation is used to get effective thickness distribution and permeability distribution of each layer.

Based on this reservoir model, block water cut is fitted in the production period and data from 10 testing wells are fitted in the testing period. Finally, we can get the distributions as follows: block and single well water-cut fitting curve (Figs 12.132 and 12.133), streamline distribution at block well shut-in moment (Figs 12.134 and 12.135), pressure fitting curve of testing well (Figs 12.136 to Fig. 12.145; interpretation result is shown in Table 12.23), permeability distribution (Figs 12.146 and 12.147), residual oil saturation distribution (Figs 12.148 and 12.149) and remaining geological reserves abundance distribution (Figs 12.150 and 12.151).

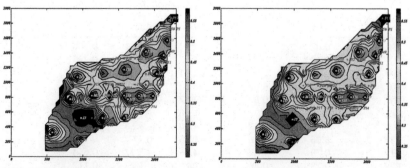

FIGURE 12.148 Oil saturation distribution of AII5 at the testing period.

FIGURE 12.149 Oil saturation distribution of AII6 at the testing period.

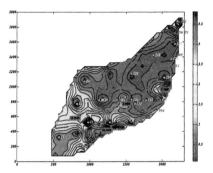

FIGURE 12.150   Remaining reserves abundance distribution of AII5 at the testing period (t/m²).

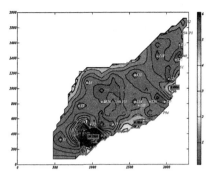

FIGURE 12.151   Remaining reserves abundance distribution of AII6 at the testing period (t/m²).

12.4. CHAPTER SUMMARY

In this chapter, streamline numerical well testing interpretation software was used to study practical field application to many types of reservoirs in different oil fields. The dynamic parameter distribution of well testing interpretation parameters and residual oil distribution in the objective area were obtained, which provide a theoretical foundation for the formulation and modification of oil field major development plans; and simultaneously, the reliability of streamline numerical well testing interpretation theory and method and the practicality of streamline numerical well testing interpretation software are illustrated.

Bibliography

Abbaszadeh, M.D. Two-phase transient testing of water injection wells in layered reservoirs. SPE 22680.

Aly, A., Chen, H.Y., Lee, W.J. A new technique for analysis of wellbore pressure from multi-layered reservoirs with unequal initial pressures to determine individual layer properties. SPE 29176.

Al-Huthali, A.H., Datta-Gupta, A. Streamline simulation of water injection in naturally fractured reservoirs. SPE 89443.

Al-Thawad, M.F., Issaka, M.B., Agyapong, D., Banerjee, R. A simple approach to numerical analysis of complex well tests. SPE 81514.

Aly, A., Lee, W. J. A new pre-production well test for analysis of multilayered commingled reservoirs with unequal initial pressures. SPE 27730.

Aly, A., Lee, W.J. Computational modeling of multi-layered reservoirs with unequal initial pressures: development of a new pre-production well test. SPE 29586.

Aly, A., Holditch, S.A., Lee, W.J. Characterizing multi-layered reservoirs using a new, simple, inexpensive and environmentally sensitive pre-production well test. SPE 36624.

Archer, A.R., Horne, N.R. The green element method for numerical well test analysis. SPE 62916.

Archer, R.A., Yildiz, T.T. Transient Well Index for Numerical Well Test Analysis. SPE 71572.

Archer, A.R., Yildiz, T.T. An integrated approach to interval pressure transient test analysis using analytical and numerical methods. SPE 81515.

Batycky, R.P., 1997. A Three-dimensional Two-phase Scale Streamline Simulator. PhD dissertation,The Department of Petroleum Engineering and The Committee on Graduate Studies, Stanford University.

Batycky, R.P., Blunt, M.J., Thiele, M.R. A 3D field scale streamline simulator with gravity and changing well conditions, SPE 36726 in Proceedings of the 1996 SPE Annual Technical Conference and Exhibition, Denver, CO, Oct 6–9.

Bidaux, P., Whittle, T.M., Coveney, P.J., Gringarten, A.C. Analysis of pressure and rate transient data from wells in multilayered reservoirs theory and application. SPE 24679.

Bourdet, D. Pressure behavior of layered reservoirs with crossflow. SPE 13628.

Bourdet, D., Gringarten, A.C., Determination of fissure volume and block size in fractured reservoirs by type-curve analysis. Society of Petroleum Engineers Annual Fall Technical Conference and Exhibit, Dallas, Texas (1980) Paper SPE9293.

Cheng, H., Osaco, I., Datta-Gupta, A. A rigorous compressible streamline formulation for two and three-phase black-oil simulation. SPE 96866.

Chengtai, G. The crossflow behavior and the determination of reservoir parameters by draw-down tests in multilayer reservoirs. SPE 12580.

Chengtai, G. The determination of total productivity by a constant pressure flow test and the cross flow behavior in multilayer reservoirs. SPE 12581.

Chengtai, G. Determination of parameters for individual layers in multilayer reservoirs by transient well tests. SPE 13952.

Coskuner, G., Ramler, B.L., Brown, W. M. Design, implementation and analysis of multilayer pressure transient tests in white rose field. SPE 63080.

Crane, M., Bratvedt, F., Bratvedt, K. A fully compositional streamline simulator. SPE 63156.

Datta-Gupta, A., King, M.J., A semianalytic approach to tracer flow modeling in heterogeneous permeable media, Adv Water Resour 18 (1995) 9–21.

Di Donaato, G., Huang, W., Blunt, M. Streamline-based dual porosity simulation of fractured reservoirs. SPE 84036.

Ehlig-Economides, A.C., Joseph, J. A new test for determination of individual layer properties in a multilayered reservoir. SPE 14167.

Gao, C. Determination of individual later properties by layer-by-layer well tests in multilayer reservoirs with crossflow. SPE 15321.

Grinestaff, G.H. Waterflood pattern allocations quantifying the injector to producer relationship with streamline simulation. SPE 54616.

Gringarten, A.C. Interpretation of tests in fissured and multilayered reservoirs with double-porosity behavior: theory and practice. SPE 10044.

Gringarten, A.C., Ramey, H.J., Use of source and greens functions in solving unsteady flow problems in reservoirs, Soc. Petrol. Eng. J 13 (1973) 285–296.

Gringarten, A.C., Ramey, H.J., Raghavan, R., Unsteady-state pressure distributions created by a well with a single infinite conductivity vertical fracture, Soc. Petrol. Eng. J 14 (1974) 247–360.

Han, D., Chen, Q., Yan, C., Reservoir Simulation Basis, (1993) Petroleum Industry Press, Beijing.

Hastings, J.J., Muggeridge, A.H., Blunt, M.J. A new streamline method for evaluating uncertainty in small-scale, two-phase flow properties. SPE 66349.

Horner, D.R., Pressure build-up in wells. Third World Petroleum Congress, May28–June6 (1950) The Hague, The Netherlands.

Ingebrigtsen, L., Bratvedt, F., Berge, J. A streamline based approach to solution of three-phase flow. SPE 51904.

Jackson, R.R., Banerjee, R. Advances in multilayer reservoir testing and analysis using numerical well testing and reservoir simulation. SPE 62917.

Jatmiko, W., Daltaban, T.S., Archer, J.S. Multi-phase flow well test analysis in multi-layer reservoirs. SPE 36557.

Jessen, K., Orr Jr., F.M. Compositional streamline simulation. SPE 77379.

Kamal, M.M., Pan, Y., Landa, J.L., et al. Numerical well testing a method to use transient testing results in reservoir simulation. SPE 95905.

Kuchuk, F.J., Shah, P.C., Ayestaran, L., Nicholson, B. Application of multilayer testing and analysis: a field case. SPE 15419.

Kucuk, F., Karakas, M., Ayestaran, L. Well testing and analysis techniques for layered reservoirs. SPE 13081.

Kulkarni, K.N., 2000. Estimating Absolute and Relative Permeability Using Dynamic Data: A Streamline Approach. PhD dissertation, The Office of Graduate Studies, Texas A&M University.

Larsen, L. Determination of skin factors and flow capacities of individual layers in two-layered reservoirs. SPE 11138.

Larsen, L. Boundary effects in pressure-transient data from layered reservoirs. SPE 19797.

Lee, W.J., Spivey, P.J. Numerical and analytical well test analysis a case history. SPE 50946.

Lefkovits, H.C., Hazebroek, P., Allen, E.E., Matthews, C.S. A study of the behavior of bounded reservoirs composed of stratified layers. SPE 1329.

Miller, C.C., Dyes, A.B., Hutchinson, C.A., The estimation of permeability and reservoir pressure from bottom hole pressure build-up characteristics, Petroleum Transactions, AIME Vol. 189 (1950) 91–104.

Moore, H.F., Wishart, H.B., An "overnight" test for detemining endurance limit, Proc. ASTM Vol. 33(pt. 11) (1933) 334–340.

Moreno, J., Kazemi, H., Gilman, J.R. Streamline simulation of countercurrent water-oil and gas-oil flow in naturally fractured dual-porosity reservoirs. SPE 89880.

Olarewaju, J.S., Lee, W.J. Pressure behavior of layered and dual-porosity reservoirs in the presence of wellbore effects. SPE 17302.

Park, H., Horne, R.N. Well test analysis of a multilayered reservoir with formation crossflow. SPE 19800.

Peddibhotla, S., Cubillos, H., Datta-Gupta, A., Wu, C.H. A rapid simulation of multiphase flow through fine-scale geostatistical realizations using a new 3-d streamline model. SPE 36008.

Pinzon, C.L., Chen, H.-Y., Teufel, L.W. Numerical well test analysis of stress-sensitive reservoirs. SPE 71034.

Portella, R.C.M., Hewett, T.A. Fast 3-D reservoir simulation and applications using streamlines. SPE 39061.

Prijambodo, R., Raghavan, R., Reynolds, A.C. Well test analysis for wells producing layered reservoirs with crossflow. SPE 10262.

Puchyr, P.J. A numerical well test model. SPE 21815.

Raghavan, R., Dixon, T.N., Robinson, S.W., Phan, V.Q. Integration of geology, geophysics and numerical simulation in the interpretation of a well test in a fluvial reservoir. SPE 62983.

Raghavan, R., Dixon, T.N., Phan, V.Q., et al. Integration of geology, geophysics, and numerical simulation in the interpretation of a well test in a fluvial reservoir. SPE 72097.

Ranney, H.J., Short-time well test data interpretation in the presence of skin effect and wellbore storage, J. Petrol. Technol 22(1) (1970) 9–104.

Ryan, T.C., Sweeney, M.J., Jamieson Jr., W.H., et al. Individual layer transient tests in low-pressure, multi-layered reservoirs. SPE 27928.

Sahni, A., Hatzignatiou, D.G. Pressure transient analysis in multilayered faulted reservoirs. SPE 29674.

Samier, P., Quettier, L., Thiele, M. Applications of streamline simulations to reservoir studies. SPE 66362.

Shah, P.C., Thambynayagam, R.K.M. Transient pressure response of a well with partial completion in a two-layer crossflowing reservoir. SPE 24681.

Spath, J., McCants, S. Waterflood optimization using a combined geostatistical-3d streamline simulation approach. SPE 38355.

Tariq, S. M., Ramey Jr., H. J. Drawdown behavior of a well with storage and skin effect communicating with layers of different radii and other characteristics. SPE 7453.

Thiele, M.R, Simulating flow in heterogeneous systems using streamtubes and streamlines, Soc. Pet. Eng. Reservoir Eng 11(1) (1996) 5–12.

Thiele, R.M., Batycky, P.R., Blunt, M.J. A streamline-based 3d field-scale compositional reservoir simulator. SPE 38889.

Thiele, M.R., Batycky, R.P., Blunt, M.J., Orr, F.M., Simulating flow in heterogeneous media using streamtubes and streamlines, SPERE, 10(1) (1966) 5–12.

Verga, F.M., Griffa, G.L., Aldegheri, A. Advanced well simulation in a multilayered reservoir. SPE 68821.

Wang, B., Lake, W.L., Pope, A.G. Development and application of a streamline micellar/polymer simulator. SPE 10290.

Wu, M., 2007. Streamline numerical welltesting interpretation method for layered reservoirs. Dongying University of Petroleum (East China).

Wu, M., Yao, J., Streamline numerical well-testing interpretation model for multilayered reservoirs, Petroleum Exploration and Development 34(5) (2007).

Wu, M., Yao, J., A streamline-based polymer flooding numerical well testing interpretation model and applications, Petroleum Drilling Techniques 35(2) (2007).

Wu, M. Yao, J., Streamline numerical well-testing interpretation model for irregularly contaminated wells with application, Xinjiang Petroleum Geology 30(3) (2009).

Wu, M. Yao, J. et al. Numerical well testing auto-matching applying double population genetic algorithm (in Chinese) Petroleum Geology and Recovery Efficiency 14(2) (2007).

Yao J., 2000. Study on Theory of Numerical Well Test. Dongying University of Petroleum (East China).

Yao, J., Wu, M., Streamline numerical well-testing interpretation model and pressure response for partially perforated wells, Petroleum Exploration and Development 36(4) (2009).

Yao, J., Wu, M., Dai, W., Zhu, Y., Streamline numerical well test interpretation model, Acta Petrolei Sinica 27(3) (2006) 96–99.

Yao, J., Wu, M., Hu, H., Streamline numerical well-testing combination flooding interpretation model for alkaline–polymer reservoirs and its application, Acta Petrolei Sinica 34(5) (2007).

Index

Printed in the United States
By Bookmasters